INVESTIGATIONS IN NUMBER, DATA, AND SPACE

Measuring

Bigger, Taller, Heavier, Smaller

Grade 1

Also appropriate for Grade 2

Tracey Wright
Jan Mokros
Susan Jo Russell

Developed at TERC, Cambridge, Massachusetts

Dale Seymour Publications®
White Plains, New York

The *Investigations* curriculum was developed at TERC (formerly Technical Education Research Centers) in collaboration with Kent State University and the State University of New York at Buffalo. The work was supported in part by National Science Foundation Grant No. ESI-9050210. TERC is a nonprofit company working to improve mathematics and science education. TERC is located at 2067 Massachusetts Avenue, Cambridge, MA 02140.

This project was supported, in part,
by the
National Science Foundation
Opinions expressed are those of the authors
and not necessarily those of the Foundation

Managing Editor: Catherine Anderson
Series Editor: Beverly Cory
ESL Consultant: Nancy Sokol Green
Production/Manufacturing Director: Janet Yearian
Production/Manufacturing Manager: Karen Edmonds
Production/Manufacturing Coordinator: Amy Changar
Design Manager: Jeff Kelly
Design: Don Taka
Composition: Andrea Reider
Illustrations: DJ Simison, Rachel Gage, Carl Yoshihara
Cover: Bay Graphics

This book is published by Dale Seymour Publications®, an imprint of Addison Wesley Longman, Inc.

> Dale Seymour Publications
> 10 Bank Street
> White Plains, NY 10602
> Customer Service: 800-872-1100

Copyright © 1998 by Dale Seymour Publications®. All rights reserved. Printed in the United States of America.

Limited reproduction permission: The publisher grants permission to individual teachers who have purchased this book to reproduce the blackline masters as needed for use with their own students. Reproduction for an entire school or school district or for commercial use is prohibited.

Order number DS43706
ISBN 1-57232-471-6
8 9 10 11 12-ML-05 04 03 02

Printed on Recycled Paper

INVESTIGATIONS IN NUMBER, DATA, AND SPACE®

T E R C

Principal Investigator   Susan Jo Russell
Co-Principal Investigator   Cornelia C. Tierney
Director of Research and Evaluation   Jan Mokros
Director of K–2 Curriculum   Karen Economopoulos

Curriculum Development
Karen Economopoulos
Marlene Kliman
Jan Mokros
Megan Murray
Susan Jo Russell
Tracey Wright

Evaluation and Assessment
Mary Berle-Carman
Jan Mokros
Andee Rubin

Teacher Support
Irene Baker
Megan Murray
Judy Storeygard
Tracey Wright

Technology Development
Michael T. Battista
Douglas H. Clements
Julie Sarama

Video Production
David A. Smith
Judy Storeygard

Administration and Production
Irene Baker
Amy Catlin

Cooperating Classrooms for This Unit
Elizabeth A. Pedrini
Arlington Public Schools
Arlington, MA

Malia Scott
Brookline Public Schools
Brookline, MA

Consultants and Advisors
Deborah Lowenberg Ball
Michael T. Battista
Marilyn Burns
Douglas H. Clements
Ann Grady

CONTENTS

About the *Investigations* Curriculum	I-1
How to Use This Book	I-2
Technology in the Curriculum	I-8
About Assessment	I-10

Bigger, Taller, Heavier, Smaller
Overview	I-12
Materials List	I-16
About the Mathematics in This Unit	I-17
About the Assessment in This Unit	I-19
Preview for the Linguistically Diverse Classroom	I-20

Investigation 1: Weighing and Balancing — 2
Sessions 1 and 2: Exploring Weight	4
Sessions 3 and 4: Comparing Two Objects	15
Sessions 5 and 6: Balancing Groups of Objects	23

Investigation 2: Filling — 32
Session 1: A Cupful of Sand	34
Sessions 2, 3 and 4: Filling Space	41
Sessions 5, 6, and 7: Comparing Containers	50

Investigation 3: Measuring Length — 60
Session 1: Longer Than, Shorter Than	62
Session 2: Measuring with Hands and Feet	68
Session 3: Feet Lengths	74
Session 4 and 5: Measuring with Cubes	78

Appendix: About Classroom Routines	87
Appendix: Vocabulary Support for Second-Language Learners	95
Blackline Masters: Family Letter, Student Sheets, Teaching Resources, Practice Pages	97

TEACHER NOTES

About Weight, Capacity, and Length	13
Working with a Partner	39
Talking About Halves	48
Pattern Block Shapes	49
Learning About Length	65
Assessment: Representing Objects in Order	82
About Choice Time	84
Keeping Track of Students' Work	86

WHERE TO START

The first-time user of *Bigger, Taller, Heavier, Smaller* should read the following:

- About the Mathematics in This Unit — I-17
- Teacher Note: About Weight, Capacity, and Length — 13
- Teacher Note: Working with a Partner — 39
- Teacher Note: Learning About Length — 65
- Teacher Note: Assessment: Representing Objects in Order — 82

When you next teach this same unit, you can begin to read more of the background. Each time you present the unit, you will learn more about how your students understand the mathematical ideas.

ABOUT THE *INVESTIGATIONS* CURRICULUM

Investigations in Number, Data, and Space® is a K–5 mathematics curriculum with four major goals:

- to offer students meaningful mathematical problems
- to emphasize depth in mathematical thinking rather than superficial exposure to a series of fragmented topics
- to communicate mathematics content and pedagogy to teachers
- to substantially expand the pool of mathematically literate students

The *Investigations* curriculum embodies an approach radically different from the traditional textbook-based curriculum. At each grade level, it consists of a set of separate units, each offering 2–8 weeks of work. These units of study are presented through investigations that involve students in the exploration of major mathematical ideas.

Approaching the mathematics content through investigations helps students develop flexibility and confidence in approaching problems, fluency in using mathematical skills and tools to solve problems, and proficiency in evaluating their solutions. Students also build a repertoire of ways to communicate about their mathematical thinking, while their enjoyment and appreciation of mathematics grow.

The investigations are carefully designed to invite all students into mathematics—girls and boys, members of diverse cultural, ethnic, and language groups, and students with different strengths and interests. Problem contexts often call on students to share experiences from their family, culture, or community. The curriculum eliminates barriers—such as work in isolation from peers, or emphasis on speed and memorization—that exclude some students from participating successfully in mathematics. The following aspects of the curriculum ensure that all students are included in significant mathematics learning:

- Students spend time exploring problems in depth.
- They find more than one solution to many of the problems they work on.
- They invent their own strategies and approaches, rather than relying on memorized procedures.
- They choose from a variety of concrete materials and appropriate technology, including calculators, as a natural part of their everyday mathematical work.
- They express their mathematical thinking through drawing, writing, and talking.
- They work in a variety of groupings—as a whole class, individually, in pairs, and in small groups.
- They move around the classroom as they explore the mathematics in their environment and talk with their peers.

While reading and other language activities are typically given a great deal of time and emphasis in elementary classrooms, mathematics often does not get the time it needs. If students are to experience mathematics in depth, they must have enough time to become engaged in real mathematical problems. We believe that a minimum of 5 hours of mathematics classroom time a week—about an hour a day—is critical at the elementary level. The plan and pacing of the *Investigations* curriculum are based on that belief.

We explain more about the pedagogy and principles that underlie these investigations in Teacher Notes throughout the units. For correlations of the curriculum to the NCTM Standards and further help in using this research-based program for teaching mathematics, see the following books:

- *Implementing the* Investigations in Number, Data, and Space® *Curriculum*
- *Beyond Arithmetic: Changing Mathematics in the Elementary Classroom* by Jan Mokros, Susan Jo Russell, and Karen Economopoulos

HOW TO USE THIS BOOK

This book is one of the curriculum units for *Investigations in Number, Data, and Space*. In addition to providing part of a complete mathematics curriculum for your students, this unit offers information to support your own professional development. You, the teacher, are the person who will make this curriculum come alive in the classroom; the book for each unit is your main support system.

Although the curriculum does not include student textbooks, reproducible sheets for student work are provided in the unit and are also available as Student Activity Booklets. Students work actively with objects and experiences in their own environment and with a variety of manipulative materials and technology, rather than with a book of instruction and problems. We strongly recommend use of the overhead projector as a way to present problems, to focus group discussion, and to help students share ideas and strategies.

Ultimately, every teacher will use these investigations in ways that make sense for his or her particular style, the particular group of students, and the constraints and supports of a particular school environment. Each unit offers information and guidance for a wide variety of situations, drawn from our collaborations with many teachers and students over many years. Our goal in this book is to help you, a professional educator, implement this curriculum in a way that will give all your students access to mathematical power.

Investigation Format

The opening two pages of each investigation help you get ready for the work that follows.

What Happens This gives a synopsis of each session or block of sessions.

Mathematical Emphasis This lists the most important ideas and processes students will encounter in this investigation.

What to Plan Ahead of Time These lists alert you to materials to gather, sheets to duplicate, transparencies to make, and anything else you need to do before starting.

INVESTIGATION 2

Filling

What Happens

Session 1: A Cupful of Sand Students experiment with how many spoonfuls of sand it takes to fill a paper or plastic cup. They discuss ways to make the filling process more consistent, repeat the experiment, and compare results.

Sessions 2, 3, and 4: Filling Space During Choice Time, students do two activities that involve filling two- and three-dimensional space. In Which Holds More Sand? they compare the capacity of two containers. In Block Puzzles, they explore ideas of area as they use pattern blocks to fill a shape outline exactly. Those ready for more challenge make their own Block Puzzles by identifying sets of pattern blocks that do and do not fit a particular outline.

Sessions 5, 6, and 7: Comparing Containers Choice Time continues for three more sessions as students continue working on Block Puzzles and two new activities that involve filling containers. In Comparing Bottles, students look for two bottles that hold the same amount of water. In Which Holds More Cubes? (a Teacher Checkpoint), students compare the number of cubes that different containers can hold. Students ready for more challenge can order the containers according to how many cubes each holds. The sessions conclude with a group experiment in which the class figures out which two of three containers hold the same amount of water.

Routines Refer to the section About Classroom Routines (pp. 87–94) for suggestions on integrating into the school day regular practice of mathematical skills in counting, exploring data, and understanding time and changes.

Mathemathical Emphasis

- Developing language to describe and compare capacity
- Comparing capacities
- Measuring and comparing capacity using nonstandard units
- Collecting and keeping track of data

INVESTIGATION 2

What to Plan Ahead of Time

Materials

- Plastic teaspoons: 1 per student and extras (Session 1)
- Small plastic cups (4–6 ounces): at least 2 per student (Sessions 1–3)
- Play sand in large plastic tubs: about 2 cupfuls per student (Sessions 1–4)
- Leveling tools (pencils, plastic knives, rulers): 1 per pair (Sessions 1–4)
- Newspaper to cover working surfaces (Sessions 1–4, optional)
- Pattern blocks: 1 bucket per 6–8 students; paper pattern blocks, pattern block stickers, or crayons, optional (Sessions 2–7)
- Interlocking cubes: class set, available in sets of 50–60 (Sessions 5–7)
- Wide-mouthed containers (e.g., coffee cans, small plastic buckets) holding 4–18 cups of sand: 2 per sand station (Sessions 2–4)
- See-through plastic bottles (e.g., soda, water, cooking oil, dishwashing liquid): 5 per water station, including two 1-liter bottles per station (Sessions 5–7)
- Empty containers (e.g., plastic cups, margarine tubs, soup cans, coffee cans, yogurt or cottage cheese tubs) holding 10–50 interlocking cubes: 20–24 for the class (Sessions 5–7)
- Water buckets, funnels: 1 of each per water station (Sessions 5–7)
- Food coloring (Sessions 5–7, optional)
- Three different-shaped containers, two having the same capacity (Session 7)
- Chart paper, unlined paper (available)

Other Preparation

- Before Session 2, set up two or three sand stations, each with a large tub of sand, two containers of different capacities, a plastic cup, and a leveling tool. Label containers with different symbols (e.g., star, heart). Tape down newspaper to protect surfaces.
- Before Session 5, set up two or three water stations, each with a bucket of water, a funnel, a plastic cup, two 1-liter plastic bottles of different shapes, and three other bottles. Adding a few drops of food coloring makes the water easier to see.
- Also before Session 5, label with letters the containers that hold 10–50 cubes.
- If you do not have manufactured paper pattern blocks or stickers, you can duplicate pages 112–117 on construction paper. Enlist adult help in cutting apart the shapes.
- Duplicate the following student sheets and teaching resources, located at the end of this unit. If you have Student Activity Booklets, no copying is needed.

For Sessions 2, 3, and 4
Student Sheet 5, Which Holds More? (p. 103): 2 per student (1 for Sessions 5–7)
Student Sheets 6–9, Block Puzzles A–D (p. 104): 1 per student
Student Sheets 10–11, Block Puzzles E–F (p. 108): 1 per student, optional

For Sessions 5, 6, and 7
Student Sheet 12, Comparing Bottles (p. 110): 1 per student
Student Sheet 13, Two Containers (p. 111): 1 per student, homework

---

I-2 ■ *Bigger, Taller, Heavier, Smaller*

Sessions Within an investigation, the activities are organized by class session, a session being at least a one-hour math class. Sessions are numbered consecutively through an investigation. Often several sessions are grouped together, presenting a block of activities with a single major focus.

When you find a block of sessions presented together—for example, Sessions 1, 2, and 3—read through the entire block first to understand the overall flow and sequence of the activities. Make some preliminary decisions about how you will divide the activities into three sessions for your class, based on what you know about your students. You may need to modify your initial plans as you progress through the activities, and you may want to make notes in the margins of the pages as reminders for the next time you use the unit.

Be sure to read the Session Follow-Up section at the end of the session block to see what homework assignments and extensions are suggested as you make your initial plans.

While you may be used to a curriculum that tells you exactly what each class session should cover, we have found that the teacher is in a better position to make these decisions. Each unit is flexible and may be handled somewhat differently by every teacher. While we provide guidance for how many sessions a particular group of activities is likely to need, we want you to be active in determining an appropriate pace and the best transition points for your class. It is not unusual for a teacher to spend more or less time than is proposed for the activities.

Activities The activities include pair and small-group work, individual tasks, and whole-class discussions. In any case, students are seated together, talking and sharing ideas during all work times. Students most often work cooperatively, although each student may record work individually.

Choice Time In most units, some sessions are structured with activity choices. In these cases, students may work simultaneously on different activities focused on the same mathematical ideas. Students choose which activities they want to do, and they cycle through them. You will need to decide how to set up and introduce these activities and how to let students make their choices. Some

---

> **Session 1**
>
> # A Cupful of Sand
>
> **Materials**
> - Plastic teaspoons (1 per student and extras)
> - Paper or plastic cups (1 per student and extras)
> - Sand in tubs
> - Newspaper to cover surfaces (optional)
> - Pencils, rulers, knives for leveling
> - Chart paper
> - Unlined paper
>
> **What Happens**
> Students experiment with how many spoonfuls of sand it takes to fill a paper or plastic cup. They discuss ways to make the filling process more consistent, repeat the experiment, and compare results. Their work focuses on:
> - estimating the number of units needed to fill a container
> - experimenting with and describing techniques for filling
> - collecting and interpreting data
> - counting and keeping track
>
> **Activity**
>
> **Filling Cups: Experiment 1**
>
> Distribute cups, spoons, and tubs of sand to students seated at tables or desks. Pass out unlined paper for recording.
>
> Today we are going to do an experiment. Everyone has a cup and a spoon. All the cups are the same size, and all the spoons are the same size. Look carefully at your cup and spoon. See if you can estimate, or guess, how many spoonfuls of sand will fit in that cup.
>
> Give students some time to think about this. Suggest that they compare the size of the cup and the spoon and try to visualize the number of spoonfuls in the cup.
>
> Before we collect your estimates, you may each put two spoonfuls of sand in your cup. See if that helps you think about how many spoonfuls it will take to fill the cup completely.
>
> Allow a moment for students to think or to talk with a neighbor about their guesses. Then record guesses on chart paper, in a column format, under the heading Estimates. Collect a good sample of estimates, but you need not call on every student.
>
> What do you notice about these numbers?

---

teachers set up choices as stations around the room, while others post the list of available choices and allow students to collect their own materials and choose their own work space. You may need to experiment with a few different structures before finding a setup that works best for you.

Extensions These follow-up activities are opportunities for some or all students to explore a topic in greater depth or in a different context. They are not designed for "fast" students; mathematics is a multifaceted discipline, and different students will want to go further in different investigations. Look for and encourage the sparks of interest and enthusiasm you see in your students, and use the extensions to help them pursue these interests.

Excursions Some of the *Investigations* units include excursions—blocks of activities that could be omitted without harming the integrity of the unit. This is one way of dealing with the great depth and variety of elementary mathematics—much more than a class has time to explore in any one year. Excursions give you the flexibility to make different choices from year to year, doing the excursion in one unit this time, and next year trying another excursion.

Tips for the Linguistically Diverse Classroom At strategic points in each unit, you will find concrete suggestions for simple modifications of the teaching strategies to encourage the participation of all students. Many of these tips offer alternative ways to elicit critical thinking from students at varying levels of English proficiency, as well as from other students who find it difficult to verbalize their thinking.

The tips are supported by suggestions for specific vocabulary work to help ensure that all students can participate fully in the investigations. The Preview for the Linguistically Diverse Classroom (p. I-20) lists important words that are assumed as part of the working vocabulary of the unit. Second-language learners will need to become familiar with these words in order to understand the problems and activities they will be doing. These terms can be incorporated into students' second-language work before or during the unit. Activities that can be used to present the words are found in the appendix, Vocabulary Support for Second-Language Learners (p. 95). In addition, ideas for making connections to students' language and cultures, included on the Preview page, help the class explore the unit's concepts from a multicultural perspective.

Classroom Routines Activities in counting, exploring data, and understanding time and changes are suggested for routines in the grade 1 *Investigations* curriculum. Routines offer ongoing work with this important content as a regular part of the school day. Some routines provide more practice with content presented in the curriculum; others extend the curriculum; still others explore new content areas.

Plan to incorporate a few of the routine activities into a standard part of your daily schedule, such as morning meeting. When opportunities arise, you can also include routines as part of your work in other subject areas (for example, keeping a weather chart for science). Most routines are short and can be done whenever you have a spare 10–15 minutes, such as before lunch or recess or at the end of the day.

You will need to decide how often to present routines, what variations are appropriate for your class, and at what points in the day or week you will include them. A reminder about classroom routines is included on the first page of each investigation. Whatever routines you choose, your students will gain the most from these routines if they work with them regularly.

Materials

A complete list of the materials needed for teaching this unit is found on p. I-16. Some of these materials are available in kits for the *Investigations* curriculum. Individual items can also be purchased from school supply dealers.

Classroom Materials In an active mathematics classroom, certain basic materials should be available at all times: interlocking cubes, pencils, unlined paper, graph paper, calculators, and things to count with. Some activities in this curriculum require scissors and glue sticks or tape. Stick-on notes and large paper are also useful materials

throughout. So that students can independently get what they need at any time, they should know where these materials are kept, how they are stored, and how they are to be returned to the storage area. Many teachers have found that stopping 5 minutes before the end of each session so that students can finish their work and clean up is helpful in maintaining classroom materials. You'll find that establishing such routines at the beginning of the year is well worth the time and effort.

Technology Calculators are introduced to students in the first unit of the grade 1 sequence, *Mathematical Thinking at Grade 1*. By freely exploring and experimenting, students become familiar with this important mathematical tool.

Computer activities at grade 1 use a software program, called *Shapes,* that was developed especially for the *Investigations* curriculum. This program is introduced in the geometry unit, *Quilt Squares and Block Towns.* Using *Shapes,* students explore two-dimensional geometry while making pictures and designs with pattern block shapes and tangram pieces.

Although the software is linked to activities only in the geometry unit, we recommend that students use it throughout the year. Thus, you may want to introduce it when you introduce pattern blocks in *Mathematical Thinking at Grade 1.* How you use the computer activities depends on the number of computers you have available. Suggestions are offered in the geometry unit for how to organize different types of computer environments.

Children's Literature Each unit offers a list of suggested children's literature (p. I-16) that can be used to support the mathematical ideas in the unit. Sometimes an activity is based on a specific children's book, with suggestions for substitutions where practical. While such activities can be adapted and taught without the book, the literature offers a rich introduction and should be used whenever possible.

Student Sheets and Teaching Resources Student recording sheets and other teaching tools needed for both class and homework are provided as reproducible blackline masters at the end of each unit. They are also available as Student Activity Booklets. These booklets contain all the sheets each student will need for individual work, freeing you from extensive copying (although you may need or want to copy the occasional teaching resource on transparency film or card stock, or make extra copies of a student sheet).

We think it's important that students find their own ways of organizing and recording their work. They need to learn how to explain their thinking with both drawings and written words, and how to organize their results so someone else can understand them. For this reason, we deliberately do not provide student sheets for every activity. Regardless of the form in which students do their work, we recommend that they keep a mathematics notebook or folder so that their work is always available for reference.

Homework In *Investigations,* homework is an extension of classroom work. Sometimes it offers review and practice of work done in class, sometimes preparation for upcoming activities, and sometimes numerical practice that revisits work in earlier units. Homework plays a role both in supporting students' learning and in helping inform families about the ways in which students in this curriculum work with mathematical ideas.

Depending on your school's homework policies and your own judgment, you may want to assign more homework than is suggested in the units. For this purpose you might use the practice pages, included as blackline masters at the end of this unit, to give students additional work with numbers.

For some homework assignments, you will want to adapt the activity to meet the needs of a variety of students in your class: those with special needs, those ready for more challenge, and second-language learners. You might change the numbers in a problem, make the activity more or less complex, or go through a sample activity with those who need extra help. You can modify any student sheet for either homework or class use. In particular, making numbers in a problem smaller or larger can make the same basic activity appropriate for a wider range of students.

Another issue to consider is how to handle the homework that students bring back to class—how to recognize the work they have done at home without spending too much time on it. Some teachers hold a short group discussion of different approaches to the assignment; others ask students to share and discuss their work with a neighbor, or post the homework around the room and give students time to tour it briefly. If you want to keep track of homework students bring in, be sure it ends up in a designated place.

***Investigations* at Home** It is a good idea to make your policy on homework explicit to both students and their families when you begin teaching with *Investigations*. How frequently will you be assigning homework? When do you expect homework to be completed and brought back to school? What are your goals in assigning homework? How independent should families expect their children to be? What should the parent or guardian's role be? The more explicit you can be about your expectations, the better the homework experience will be for everyone.

Investigations at Home (a booklet available separately for each unit, to send home with students) gives you a way to communicate with families about the work students are doing in class. This booklet includes a brief description of every session, a list of the mathematics content emphasized in each investigation, and a discussion of each homework assignment to help families more effectively support their children. Whether or not you are using the *Investigations* at Home booklets, we expect you to make your own choices about homework assignments. Feel free to omit any and to add extra ones you think are appropriate.

Family Letter A letter that you can send home to students' families is included with the blackline masters for each unit. Families need to be informed about the mathematics work in your classroom; they should be encouraged to participate in and support their children's work. A reminder to send home the letter for each unit appears in one of the early investigations. These letters are also available separately in Spanish, Vietnamese, Cantonese, Hmong, and Cambodian.

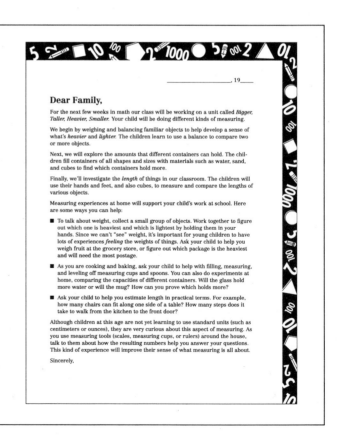

Help for You, the Teacher

Because we believe strongly that a new curriculum must help teachers think in new ways about mathematics and about their students' mathematical thinking processes, we have included a great deal of material to help you learn more about both.

About the Mathematics in This Unit This introductory section (p. I-17) summarizes the critical information about the mathematics you will be teaching. It describes the unit's central mathematical ideas and how students will encounter them through the unit's activities.

Teacher Notes These reference notes provide practical information about the mathematics you are teaching and about our experience with how students learn. Many of the notes were written in response to actual questions from teachers, or to discuss important things we saw happening in the field-test classrooms. Some teachers like to read them all before starting the unit, then review them as they come up in particular investigations.

Dialogue Boxes Sample dialogues demonstrate how students typically express their mathematical ideas, what issues and confusions arise in their thinking, and how some teachers have guided class discussions. These dialogues are based on the extensive classroom testing of this curriculum; many are word-for-word transcriptions of recorded class discussions. They are not always easy reading; sometimes it may take some effort to unravel what the students are trying to say. But this is the value of these dialogues; they offer good clues to how your students may develop and express their approaches and strategies, helping you prepare for your own class discussions.

Where to Start You may not have time to read everything the first time you use this unit. As a first-time user, you will likely focus on understanding the activities and working them out with your students. Read completely through each investigation before starting to present it. Also read those sections listed in the Contents under the heading Where to Start (p. vi).

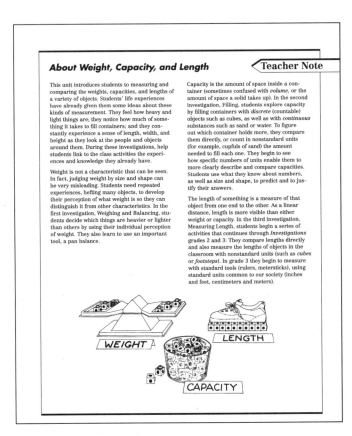

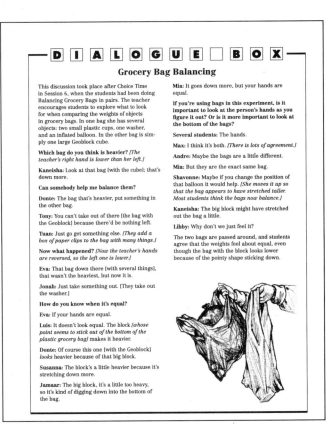

TECHNOLOGY IN THE CURRICULUM

The *Investigations* curriculum incorporates the use of two forms of technology in the classroom: calculators and computers. Calculators are assumed to be standard classroom materials, available for student use in any unit. Computers are explicitly linked to one or more units at each grade level; they are used with the unit on 2-D geometry unit at each grade, as well as with some of the units on measuring, data, and changes.

Using Calculators

In this curriculum, calculators are considered tools for doing mathematics, similar to pattern blocks or interlocking cubes. Just as with other tools, students must learn both *how* to use calculators correctly and *when* they are appropriate to use. This knowledge is crucial for daily life, as calculators are now a standard way of handling numerical operations, both at work and at home.

Using a calculator correctly is not a simple task; it depends on a good knowledge of the four operations and of the number system, so that students can select suitable calculations and also determine what a reasonable result would be. These skills are the basis of any work with numbers, whether or not a calculator is involved.

Unfortunately, calculators are often seen as tools to check computations with, as if other methods are somehow more fallible. Students need to understand that any computational method can be used to check any other; it's just as easy to make a mistake on the calculator as it is to make a mistake on paper or with mental arithmetic. Throughout this curriculum, we encourage students to solve computation problems in more than one way in order to double-check their accuracy. We present mental arithmetic, paper-and-pencil computation, and calculators as three possible approaches.

In this curriculum we also recognize that, despite their importance, calculators are not always appropriate in mathematics instruction. Like any tools, calculators are useful for some tasks, but not for others. You will need to make decisions about when to allow students access to calculators and when to ask that they solve problems without them, so that they can concentrate on other tools and skills. At times when calculators are or are not appropriate for a particular activity, we make specific recommendations. Help your students develop their own sense of which problems they can tackle with their own reasoning and which ones might be better solved with a combination of their own reasoning and the calculator.

Managing calculators in your classroom so that they are a tool, and not a distraction, requires some planning. When calculators are first introduced, students often want to use them for everything, even problems that can be solved quite simply by other methods. However, once the novelty wears off, students are just as interested in developing their own strategies, especially when these strategies are emphasized and valued in the classroom. Over time, students will come to recognize the ease and value of solving problems mentally, with paper and pencil, or with manipulatives, while also understanding the power of the calculator to facilitate work with larger numbers.

Experience shows that if calculators are available only occasionally, students become excited and distracted when they are permitted to use them. They focus on the tool rather than on the mathematics. In order to learn when calculators are appropriate and when they are not, students must have easy access to them and use them routinely in their work.

If you have a calculator for each student, and if you think your students can accept the responsibility, you might allow them to keep their calculators with the rest of their individual materials, at least for the first few weeks of school. Alternatively, you might store them in boxes on a shelf, number each calculator, and assign a corresponding number to each student. This system can give students a sense of ownership while also helping you keep track of the calculators.

Using Computers

Students can use computers to approach and visualize mathematical situations in new ways. The computer allows students to construct and manipulate geometric shapes, see objects move according to rules they specify, and turn, flip, and repeat a pattern.

This curriculum calls for computers in units where they are a particularly effective tool for learning mathematics content. One unit on 2-D geometry at each of the grades 3–5 includes a core of activities that rely on access to computers, either in the classroom or in a lab. Other units on geometry, measurement, data, and changes include computer activities, but can be taught without them. In these units, however, students' experience is greatly enhanced by computer use.

The following list outlines the recommended use of computers in this curriculum:

Grade 1
Unit: *Survey Questions and Secret Rules*
  (Collecting and Sorting Data)
Software: Tabletop, Jr.
Source: Broderbund

Unit: *Quilt Squares and Block Towns*
  (2-D and 3-D Geometry)
Software: *Shapes*
Source: provided with the unit

Grade 2
Unit: *Mathematical Thinking at Grade 2*
  (Introduction)
Software: *Shapes*
Source: provided with the unit

Unit: *Shapes, Halves, and Symmetry*
  (Geometry and Fractions)
Software: *Shapes*
Source: provided with the unit

Unit: *How Long? How Far?* (Measuring)
Software: *Geo-Logo*
Source: provided with the unit

Grade 3
Unit: *Flips, Turns, and Area* (2-D Geometry)
Software: *Tumbling Tetrominoes*
Source: provided with the unit

Unit: *Turtle Paths* (2-D Geometry)
Software: *Geo-Logo*
Source: provided with the unit

Grade 4
Unit: *Sunken Ships and Grid Patterns*
  (2-D Geometry)
Software: *Geo-Logo*
Source: provided with the unit

Grade 5
Unit: *Picturing Polygons* (2-D Geometry)
Software: *Geo-Logo*
Source: provided with the unit

Unit: *Patterns of Change* (Tables and Graphs)
Software: *Trips*
Source: provided with the unit

Unit: *Data: Kids, Cats, and Ads* (Statistics)
Software: Tabletop, Sr.
Source: Broderbund

The software provided with the *Investigations* units uses the power of the computer to help students explore mathematical ideas and relationships that cannot be explored in the same way with physical materials. With the *Shapes* (grades 1–2) and *Tumbling Tetrominoes* (grade 3) software, students explore symmetry, pattern, rotation and reflection, area, and characteristics of 2-D shapes. With the *Geo-Logo* software (grades 3–5), students investigate rotations and reflections, coordinate geometry, the properties of 2-D shapes, and angles. The *Trips* software (grade 5) is a mathematical exploration of motion in which students run experiments and interpret data presented in graphs and tables.

We suggest that students work in pairs on the computer; this not only maximizes computer resources but also encourages students to consult, monitor, and teach one another. Generally, more than two students at one computer find it difficult to share. Managing access to computers is an issue for every classroom. The curriculum gives you explicit support for setting up a system. The units are structured on the assumption that you have enough computers for half your students to work on the machines in pairs at one time. If you do not have access to that many computers, suggestions are made for structuring class time to use the unit with five to eight computers, or even with fewer than five.

ABOUT ASSESSMENT

Assessment plays a critical role in teaching and learning, and it is an integral part of the *Investigations* curriculum. For a teacher using these units, assessment is an ongoing process. You observe students' discussions and explanations of their strategies on a daily basis and examine their work as it evolves. While students are busy recording and representing their work, working on projects, sharing with partners, and playing mathematical games, you have many opportunities to observe their mathematical thinking. What you learn through observation guides your decisions about how to proceed. In any of the units, you will repeatedly consider questions like these:

- Do students come up with their own strategies for solving problems, or do they expect others to tell them what to do? What do their strategies reveal about their mathematical understanding?

- Do students understand that there are different strategies for solving problems? Do they articulate their strategies and try to understand other students' strategies?

- How effectively do students use materials as tools to help with their mathematical work?

- Do students have effective ideas for keeping track of and recording their work? Does keeping track of and recording their work seem difficult for them?

You will need to develop a comfortable and efficient system for recording and keeping track of your observations. Some teachers keep a clipboard handy and jot notes on a class list or on adhesive labels that are later transferred to student files. Others keep loose-leaf notebooks with a page for each student and make weekly notes about what they have observed in class.

Assessment Tools in the Unit

With the activities in each unit, you will find questions to guide your thinking while observing the students at work. You will also find two built-in assessment tools: Teacher Checkpoints and embedded Assessment activities.

Teacher Checkpoints The designated Teacher Checkpoints in each unit offer a time to "check in" with individual students, watch them at work, and ask questions that illuminate how they are thinking.

At first it may be hard to know what to look for, hard to know what kinds of questions to ask. Students may be reluctant to talk; they may not be accustomed to having the teacher ask them about their work, or they may not know how to explain their thinking. Two important ingredients of this process are asking students open-ended questions about their work and showing genuine interest in how they are approaching the task. When students see that you are interested in their thinking and are counting on them to come up with their own ways of solving problems, they may surprise you with the depth of their understanding.

Teacher Checkpoints also give you the chance to pause in the teaching sequence and reflect on how your class is doing overall. Think about whether you need to adjust your pacing: Are most students fluent with strategies for solving a particular kind of problem? Are they just starting to formulate good strategies? Or are they still struggling with how to start? Depending on what you see as the students work, you may want to spend more time on similar problems, change some of the problems to use smaller numbers, move quickly to more challenging material, modify subsequent activities for some students, work on particular ideas with a small group, or pair students who have good strategies with those who are having more difficulty.

Embedded Assessment Activities Assessment activities embedded in each unit will help you examine specific pieces of student work, figure out what it means, and provide feedback. From the students' point of view, these assessment activities are no different from any others. Each is a learning experience in and of itself, as well as an opportunity for you to gather evidence about students' mathematical understanding.

The embedded assessment activities sometimes involve writing and reflecting; at other times, a discussion or brief interaction between student and teacher; and in still other instances, the creation and explanation of a product. In most cases, the assessments require that students *show* what they did, *write* or *talk* about it, or do both. Having to explain how they worked through a problem helps students be more focused and clear in their mathematical thinking. It also helps them realize that doing mathematics is a process that may involve tentative starts, revising one's approach, taking different paths, and working through ideas.

Teachers often find the hardest part of assessment to be interpreting their students' work. We provide guidelines to help with that interpretation. If you have used a process approach to teaching writing, the assessment in *Investigations* will seem familiar. For many of the assessment activities, a Teacher Note provides examples of student work and a commentary on what it indicates about student thinking.

Documentation of Student Growth

To form an overall picture of mathematical progress, it is important to document each student's work in journals, notebooks, or portfolios. The choice is largely a matter of personal preference; some teachers have students keep a notebook or folder for each unit, while others prefer one mathematics notebook, or a portfolio of selected work for the entire year. The final activity in each *Investigations* unit, called Choosing Student Work to Save, helps you and the students select representative samples for a record of their work.

This kind of regular documentation helps you synthesize information about each student as a mathematical learner. From different pieces of evidence, you can put together the big picture. This synthesis will be invaluable in thinking about where to go next with a particular child, deciding where more work is needed, or explaining to parents (or other teachers) how a child is doing.

If you use portfolios, you need to collect a good balance of work, yet avoid being swamped with an overwhelming amount of paper. Following are some tips for effective portfolios:

- Collect a representative sample of work, including some pieces that students themselves select for inclusion in the portfolio. There should be just a few pieces for each unit, showing different kinds of work—some assignments that involve writing, as well as some that do not.

- If students do not date their work, do so yourself so that you can reconstruct the order in which pieces were done.

- Include your reflections on the work. When you are looking back over the whole year, such comments are reminders of what seemed especially interesting about a particular piece; they can also be helpful to other teachers and to parents. Older students should be encouraged to write their own reflections about their work.

Assessment Overview

There are two places to turn for a preview of the assessment opportunities in each *Investigations* unit. The Assessment Resources column in the unit Overview Chart (pp. I-13–I-15) identifies the Teacher Checkpoints and Assessment activities embedded in each investigation, guidelines for observing the students that appear within classroom activities, and any Teacher Notes and Dialogue Boxes that explain what to look for and what types of student responses you might expect to see in your classroom. Additionally, the section About the Assessment in This Unit (p. I-19) gives you a detailed list of questions for each investigation, keyed to the mathematical emphases, to help you observe student growth.

Depending on your situation, you may want to provide additional assessment opportunities. Most of the investigations lend themselves to more frequent assessment, simply by having students do more writing and recording while they are working.

OVERVIEW

Bigger, Taller, Heavier, Smaller

Content of This Unit To become familiar with the process of measuring, students compare weights, fill containers, and figure out the lengths of objects. They begin by weighing and balancing a variety of familiar objects to develop a sense of what's heavier and lighter. They learn to use a balance to compare the weights of two or more objects. Next students explore capacity—the amount that a container can hold—by filling containers of many different shapes and sizes with water, sand, or cubes. Finally, they investigate the measure of length, using hands, feet, and cubes to measure and compare the lengths of objects in the classroom, as well as comparing objects directly to determine which is longer. Through these investigations, students develop language to describe weight, capacity, and length; they explore qualitative comparison; and they learn to use units to measure and compare.

Connections with Other Units If you are doing the full-year *Investigations* curriculum in the suggested sequence for grade 1, this is the last of six units. The work provides a foundation for the measuring units in grade 2 *(How Long? How Far?)* and grade 3 *(From Paces to Feet.)*. This unit can also be used successfully at grade 2.

Investigations Curriculum ▪ Suggested Grade 1 Sequence

Mathematical Thinking at Grade 1 (Introduction)

Building Number Sense (The Number System)

Survey Questions and Secret Rules (Collecting and Sorting Data)

Quilt Squares and Block Towns (2-D and 3-D Geometry)

Number Games and Story Problems (Addition and Subtraction)

▶ *Bigger, Taller, Heavier, Smaller* (Measuring)

Investigation 1 ■ Weighing and Balancing

Class Sessions	Activities	Pacing
Sessions 1 and 2 (p. 4) EXPLORING WEIGHT	Introducing Which Is Heavier? Introducing the Balance Choice Time Who Sank the Boat? Homework: Something to Weigh Extension: Weight Book	minimum 2 hr
Sessions 3 and 4 (p. 15) COMPARING TWO OBJECTS	Using a Balance Introducing Balance Comparisons Teacher Checkpoint: Choice Time Extension: Balancing on a Seesaw	minimum 2 hr
Sessions 5 and 6 (p. 23) BALANCING GROUPS OF OBJECTS	Introducing Balancing Grocery Bags Introducing Balancing Balances Choice Time How Can We Tell If It's Balanced? Homework: Finding Things That Balance Extension: How Many Washers? Extension: What Balances Six Washers?	minimum 2 hr

Classroom Routines (see pp. 87–94)

Mathematical Emphasis

- Developing a sense of heavier and lighter by feel
- Developing language to describe and compare weights
- Learning to use a balance
- Comparing the weights of objects using a balance
- Representing the results of weight comparisons

Assessment Resources

Observing the Students (pp. 10, 19, 27)

Teacher Checkpoint: Choice Time—Balance Comparisons (p. 18)

Balancing Pattern Blocks (Dialogue Box, p. 30)

Materials

Balances
Small objects to weigh (tiles, pattern blocks, washers)
Plastic bags with handles
Miscellaneous classroom objects
Colored stickers
Chart paper
Unlined paper
Who Sank the Boat? (opt.)
Math Counts: Weight (opt.)
Just a Little Bit (opt.)
Student Sheets 1–4

Investigation 2 ■ Filling

Class Sessions	Activities	Pacing
Session 1 (p. 34) A CUPFUL OF SAND	Filling Cups: Experiment 1 What Is a Cupful of Sand? Filling Cups: Experiment 2 Extension: Big Spoon, Little Spoon	minimum 1 hr
Sessions 2, 3, and 4 (p. 41) FILLING SPACE	Introducing Which Holds More Sand? Introducing Block Puzzles Choice Time Discussing Pairs of Containers Extension: Capacity Book Extension: Creating Block Puzzles	minimum 3 hr
Sessions 5, 6, and 7 (p. 50) COMPARING CONTAINERS	Introducing Comparing Bottles Introducing Which Holds More Cubes? Teacher Checkpoint: Choice Time Discussing Our Work with Bottles of Water Homework: Two Containers	minimum 3 hr

Classroom Routines (see pp. 87–94)

Mathematical Emphasis

- Developing language to describe and compare capacity
- Comparing capacities
- Measuring and comparing capacity using nonstandard units
- Collecting and keeping track of data

Assessment Resources

Working with a Partner (Teacher Note, p. 39)

Observing the Students (p. 38, 45, 55)

Teacher Checkpoint: Choice Time—Which Holds More Cubes? (p. 53)

Comparing Cupfuls (Dialogue Box, p. 58)

Materials

Plastic cups
Plastic teaspoons, and larger plastic spoons
Plastic knives or rulers (opt.)
Play sand
Plastic tubs
Empty containers
Plastic bottles
Funnels
Water buckets
Pattern blocks
Pattern block cutouts and glue, or crayons
Interlocking cubes
Newspaper (opt.)
Food coloring (opt.)
Smocks (opt.)
Chart paper
Unlined paper
Student Sheets 5–13

Investigation 3 ■ Measuring Length

Class Sessions	Activities	Pacing
Session 1 (p. 62) LONGER THAN, SHORTER THAN	Which Is Longer? Longer and Shorter Homework: Foot Outlines Extension: More Pencil Comparisons	minimum 1 hr
Session 2 (p. 68) MEASURING WITH HANDS AND FEET	Measuring a Strip of Tape Measuring Things in the Classroom Homework: Shorter Than My Arm	minimum 1 hr
Session 3 (p. 74) FEET LENGTHS	Putting Feet in Order Foot Match-Ups Homework: Measuring with Hands and Feet Extension: Length Book Extension: More About Halves	minimum 1 hr
Sessions 4 and 5 (p. 78) MEASURING WITH CUBES	Measuring Objects with Cubes Assessment: Representing Objects in Order Sharing Representations Choosing Student Work to Save	minimum 2 hr

Classroom Routines (see pp. 87–94)

Mathematical Emphasis

- Developing language to describe and compare lengths
- Comparing lengths directly
- Measuring and comparing length using nonstandard units
- Ordering lengths
- Representing measurements with numbers, concrete materials, and pictures

Assessment Resources

Observing the Students (pp. 64, 75, 79)

Learning About Length (Teacher Note, p. 65)

More Steps or Less? (Dialogue Box, p. 73)

Assessment: Representing Objects in Order (p. 79 and Teacher Note, p. 82)

Materials

Miscellaneous objects to measure
Interlocking cubes
Masking tape
Chart paper
Large paper (11 by 17 inches)
Glue
Markers or crayons
Student Sheets 14–19
Teaching resource sheets

Overview ■ **I-15**

MATERIALS LIST

Following are the basic materials needed for the activities in this unit. Many of the items can be purchased from the publisher, either individually or in the Teacher Resource Package and the Student Materials Kit for grade 1. Detailed information is available on the *Investigations* order form. To obtain this form, call toll-free 1-800-872-1100 and ask for a Dale Seymour customer service representative.

Balances (at least 6 for the class)

Small identical objects, such as square tiles, hexagon pattern blocks, washers (50–60 objects, of 2–3 types, at each of six balance stations)

Pattern blocks (1 bucket per 6–8 students)

Paper pattern blocks or stickers

Interlocking cubes (class set of 1000)

Who Sank the Boat? by Pamela Allen (optional)

Funnels (2–3 for the class)

Play sand (about 2 cupfuls per student). Cat litter (the non-clumping kind) or birdseed can be used as alternatives to both sand and water.

Large plastic tubs for sand and water (4–6 for the class)

Plastic spoons (1 per student and extras)

Plastic or paper cups, 4–6 oz (at least 2 per student)

Plastic knives, rulers, or pencils for leveling (1 per pair)

See-through plastic bottles, such as bottles for soda or water, cooking oil, or cleaning substances (10–15 for the class, including 4–6 one-liter bottles)

Nonbreakable, wide-mouthed containers that hold from 4 to 18 cups of sand, such as coffee cans, small plastic buckets, plastic food containers (4–6 for the class)

Nonbreakable containers that hold from 10 to 50 interlocking cubes, such as large and small margarine tubs, paper or plastic cups, soup cans, yogurt and cottage cheese containers (about 24 for the class)

Plastic grocery bags with handles (1 per student and extras)

Common objects for weighing and measuring (pencils, crayons, boxes of paper clips, pads of stick-on notes, decks of cards, small books, cassette tapes, calculators, glue containers, boxes of staples, rolls of tape, and so forth)

Colored stickers (to mark object collections)

Newspaper (optional, to cover desks)

Smocks for water activities (optional)

Food coloring (optional)

Masking tape or colored electrical tape (1 roll)

Glue

Pencils for comparing lengths (1 per student)

Large paper (11 by 17 inches)

Chart paper

Unlined paper

Markers or crayons

The following materials are provided at the end of this unit as blackline masters. A Student Activity Booklet containing all student sheets and teaching resources needed for individual work is available.

Family Letter (p. 98)

Student Sheets 1–19 (p. 99)

Teaching Resources:
    Pattern Block Cutouts (p. 112)
    Foot Outlines (p. 124)
    100 Chart (p. 126)

Practice Pages (p. 127)

Related Children's Literature

Allen, Pamela. *Who Sank the Boat?* New York: Sandcastle Books, Putnam and Grosset, Coward-McCann, Inc. 1982.

Pluckrose, Henry. *Math Counts: Capacity.* Chicago: Childrens Press, 1995.

Pluckrose, Henry. *Math Counts: Length.* Chicago: Childrens Press, 1995.

Pluckrose, Henry. *Math Counts: Weight.* Chicago: Childrens Press, 1995.

Pomerantz, Charlotte. *The Half-Birthday Party.* New York: Clarion Books, 1984.

Tompert, Ann. *Just a Little Bit.* Boston: Houghton Mifflin, 1993.

ABOUT THE MATHEMATICS IN THIS UNIT

Young students' ideas about measuring grow out of a great deal of experience with informal measuring. This unit provides such experiences with weight, capacity, and length. The focus in first grade is primarily on the *process* of measuring, rather than on the numerical results of that process.

Comparing is a natural way for students to approach measuring. Even very young children spontaneously try to see who or what is bigger, taller, heavier, or smaller. Who is taller? Whose sand bucket holds more? In this unit, students compare such amounts both qualitatively and quantitatively.

Qualitative comparison is essential as a foundation for developing ideas about length, weight, and capacity. In this unit students compare two objects directly, for example by putting them next to each other to compare their length, or by holding one in each hand to compare their weight. As students compare objects directly, they are working on a key idea about measuring—that a single characteristic of an object may not tell you everything about its measure. In other words: You can't always tell just by looking. Sometimes large objects are heavier than small ones; a teacher usually weighs more than a first grade student. However, focusing only on size or height to determine weight can be misleading: a large balloon weighs less than a small rock.

The same is true of capacity; looks can be deceiving. A taller container holds more than a shorter one if the other dimensions of the two containers are about the same. However, a taller container might hold less than a shorter one if the shorter one is considerably wider.

Even in the measure of length, more than one characteristic is important to consider. Comparing the end points of two objects might indicate which is longer, but only if the *beginning* points of the objects are aligned. Research has documented that first graders are still developing ideas about where to focus in order to assess which is longer, which is heavier, or which holds more. Expect that many first graders will focus on only one characteristic of the objects they are considering.

Besides working with length, capacity, and weight, students will also continue with work on filling outlines with pattern blocks, introduced in the unit *Quilt Squares and Block Towns*. Through this informal introduction to area, students experiment with using different combinations of shapes to fill the same area.

In this unit students will also learn to quantify their measures of capacity and length using nonstandard units (such as "3 cupfuls of sand" or "8 cubes long"). By counting cupfuls of sand to measure capacity or counting cubes to measure a length, students are learning how to establish a unit that allows them to compare objects more precisely. Now, in addition to knowing which is longer or which holds more, students can use their knowledge of number as they make comparisons. This builds toward *Investigations* work at grades 2 and 3, when students determine *how much longer* or *how much more*.

As students measure with nonstandard units like "spoonfuls" or "hands," they are finding ways to describe capacity and length more exactly and are beginning to face the need for standard units and standard measuring techniques if they are to successfully share information. Standard units and measuring tools are introduced in *Investigations* grade 3, *From Paces to Feet*.

Developing language about measuring and comparing is another emphasis of this unit. Students need to become comfortable using their own language to describe their measuring activities, while hearing and using words that describe and compare a variety of measures: *long, short, wide, tall, heavy, light, even, full, empty, deep, narrow,* and the comparative forms of all these—*longer, wider, heavier,* and so forth.

As students describe and justify the strategies they use for measuring and comparing objects, you may see a wide range of strategies for solving measuring problems. Keep in mind that trial-and-error is an age-appropriate strategy and may be the predominant one your students use.

A final thought for teachers to keep in mind is that all measurement is approximate. This idea makes measuring quite different from counting. When we think about the number of a group of objects, we can be quite confident that there is an exact answer to be found (even if we are not able to find it). This is not the case with measurement. We might measure something to our satisfaction and say that it is 6 inches long. However, is it exactly 6 inches? If we had a measuring tool with finer markings, perhaps we would see that it is closer to $6\frac{1}{16}$ inches long. But is it exactly this long? What if we had yet a more finely calibrated measuring tool and could look through a magnifying glass? Would we see that, in fact, the length is a little under $6\frac{1}{16}$ inches?

Measuring is dependent on the sensitivity of the measuring tool, our techniques of measuring, and the accuracy of our observations. Students encounter this idea as they decide what a spoonful is, or what *full* means, or what to call a measurement of length that is more than 6 cubes but less than 7 cubes. The idea that all measurement is accurate only to a certain degree is complex, and most students will not be thinking deeply about it. However, be aware of this basic idea about measuring as you observe and listen to your students.

Mathematical Emphasis At the beginning of each investigation, the Mathematical Emphasis section tells you what is most important for students to learn about during that investigation. Many of these understandings and processes are difficult and complex. Students gradually learn more and more about each idea over many years of schooling. Individual students will begin and end the unit with different levels of knowledge and skill, but all will explore qualitative comparisons, use non-standard units to measure and compare quantitatively, and develop language to describe weight, capacity, and length.

ABOUT THE ASSESSMENT IN THIS UNIT

Throughout the *Investigations* curriculum, there are many opportunities for ongoing daily assessment as you observe, listen to, and interact with students at work. In this unit you will find two Teacher Checkpoints:

> Investigation 1, Sessions 3–4:
> Choice Time—Balance Comparisons (p. 18)
>
> Investigation 2, Sessions 5, 6, and 7:
> Choice Time—Which Holds More Cubes? (p. 53)

This unit also has one embedded assessment activity:

> Investigation 3, Session 4–5:
> Representing Objects in Order (p. 79)

In addition, you can use almost any activity in this unit to assess your students' needs and strengths. Following are questions to help you focus your observations in each investigation. You may want to keep track of your observations for each student to help you plan your curriculum and monitor students' growth.

Investigation 1: Weighing and Balancing

- How are students predicting which objects are heavier or lighter? How do they justify their reasoning? What information are they using to make their decisions? (Are they using their own perception of weight? previous experiences with objects?) Are they changing what information they use as they become more experienced?

- Can students describe aspects of weighing and balancing in their own words? Do they use words such as *heavier, lighter, balanced, weighs more, weighs less,* and so forth?

- How are students using a balance? Do they know that the down position on the balance indicates a heavier object? Do they know what the balance looks like when it is balanced? What questions do they raise as they use a balance?

- Can students use the balance to compare the weights of two objects? more than two objects? What strategies do they use to balance two groups of objects?

- How are students representing the results of weight comparisons? Do they use objects, pictures, numbers? Are their representations clear to others? How are they organized?

Investigation 2: Filling

- Can students describe in their own words situations about filling containers? Do they use words such as *full, empty, almost full, wider, shorter, taller,* and so forth to describe and compare capacities?

- Can students compare capacities directly by pouring something from one container to another? Can they compare capacities by using nonstandard units? Do they pay attention to creating their units (such as cupfuls) in a standard way? Are they able to use what they know about numbers to compare capacities?

- What methods are students using for keeping track as they fill containers? Are they finding ways to work efficiently with a partner?

- What strategies do students use to predict and compare which containers hold more? Are they relating size and shape to capacity? How do they justify which container holds more?

Investigation 3: Measuring Length

- Can students describe situations about measuring length in their own words? Do they use words such as *longer, shorter, farther, bigger, taller,* and so forth?

- Can students find which of two objects is longer by comparing them directly? How are they aligning the ends of objects to compare them?

- How do students repeat units to measure length? Do they pay attention to overlapping? to gaps? to where to start and stop? Do they use what they know about numbers as they quantify length? Can they use nonstandard units to compare lengths?

- Can students order a small group of objects by length? What strategies do they use? How do they justify their solutions?

- How are students representing their measurements? Do they use objects, pictures, numbers? Are their representations clear to others? How are they organized?

PREVIEW FOR THE LINGUISTICALLY DIVERSE CLASSROOM

In the *Investigations* curriculum, mathematical vocabulary is introduced naturally during the activities. We don't ask students to learn definitions of new terms; rather, they come to understand such words as *triangle, add, compare, data,* and *graph* by hearing them used frequently in discussion as they investigate new concepts. This approach is compatible with current theories of second-language acquisition, which emphasize the use of new vocabulary in meaningful contexts while students are actively involved with objects, pictures, and physical movement.

Listed below are some key words used in this unit that will not be new to most English speakers at this age level, but may be unfamiliar to students with limited English proficiency. You will want to spend additional time working on these words with your students who are learning English. If your students are working with a second-language teacher, you might enlist your colleague's aid in familiarizing students with these words, before and during this unit. In the classroom, look for opportunities for students to hear and use these words. Activities you can use to present the words are given in the appendix, Vocabulary Support for Second-Language Learners (p. 95).

Note: While all students will be developing mathematical meaning for the following terms through the activities in this unit, second-language students will benefit from some prior exposure to help them comprehend student-generated discussion as well as offer their own ideas related to measurement.

heavy, light, heavier, lighter, same In the first investigation, students use these terms as they explore comparative weights.

container, empty, fill, holds more, holds less, same amount In the second investigation, students explore and compare the capacity of different containers by filling them with sand and water.

size, wide, tall, narrow Students discover how different aspects of size are related to how much a container holds.

long, longer, longest, short, shorter, shortest In the third investigation, Measuring Length, students use these terms as they explore and compare the length of classroom objects and people's feet.

Multicultural Extensions for All Students

Whenever possible, encourage students to share words, objects, customs, or any aspects of daily life from their own cultures and backgrounds that are relevant to the activities in this unit. For example, extend the Grocery Bag Balancing activity in Investigation 1 to use actual nonperishable groceries. Ask families to lend packaged foods or other items that are especially representative of their cultures.

Investigations

INVESTIGATION 1

Weighing and Balancing

What Happens

Sessions 1 and 2: Exploring Weight Students begin their exploration of weight by hefting objects in their two hands and using a pan balance for comparison. As a class they read a book related to weight and become familiar with the terms *heavier, lighter,* and *balanced.*

Sessions 3 and 4: Comparing Two Objects Students talk about their use of balances in Sessions 1 and 2 and consider any issues that have arisen. They identify pairs of objects that they found difficult to compare by hefting and consider how the balance helps them compare these weights. During Choice Time, students continue hefting objects to determine their comparative weights, checking their results with a balance or against classmates' perceptions. They also use a balance to find objects that weigh *more than, less than,* and *the same as* a particular object they have chosen. Each student draws a picture showing how the balance looks for one of these comparisons.

Sessions 5 and 6: Balancing Groups of Objects Students work with partners on two Choice Time activities. In Balancing Grocery Bags, they use their developing sense of weight to balance groups of objects evenly in two bags. In Balancing Balances, they use the balances to make two groups of objects that weigh the same.

Routines Refer to the section About Classroom Routines (pp. 87–94) for suggestions on integrating into the school day regular practice of mathematical skills in counting, exploring data, and understanding time and changes.

Mathematical Emphasis

- Developing a sense of what's heavier and lighter by feel
- Developing language to describe and compare weights
- Learning to weigh with a balance
- Comparing the weights of objects using a balance
- Representing the results of weight comparisons

Looking Ahead

Sand During Investigation 2, students will be working with sand. Students need time to explore a new material before using it in structured activities. If you have not used sand in the classroom, create a sand exploration center with cups, spoons, funnels, and plastic tubing. You might also include small figures (people or animals). Establish clear expectations about how the sand and related materials will be used and cared for. Many teachers make materials available to students during free time or before and after school. If you cannot use sand, you could substitute cat litter (the non-clumping kind) or birdseed.

Water Similarly, if your class has not had a chance to experiment with water, set up water stations where students can use funnels, bottles, cups, and water in an undirected, exploratory way before Investigation 2. Again, establish clear guidelines (always pouring the water over the tub, and so forth), but expect that some spills are inevitable.

Pattern Blocks Investigation 2 also includes work with pattern blocks. If your class has not used them recently, provide time for free exploration so students become familiar with the shapes and how they fit together.

INVESTIGATION 1

What to Plan Ahead of Time

Materials

- At least 6 balances (Sessions 1–6)
- *Who Sank the Boat?* by Pamela Allan (Sandcastle, 1982) or another book about weight (Sessions 1–2)
- Plastic grocery bags with handles, all the same size (e.g., from one supermarket): 1 per student and some extras (Sessions 5–6)
- Small identical objects, such as square tiles, pattern blocks, washers: 50–60 objects, of 2 or 3 types, at each balance station (Sessions 5–6)
- Stickers in several colors to label the different object collections
- Chart paper, unlined paper (available)

Other Preparation

- Things to Weigh by Hand: Gather about 20 easy-to-hold objects with weights up to 2 pounds (e.g., pencil, calculator, container of glue, box of markers, box of staples, child's shoe). Label each item with a sticker of the same color so that loose objects can be returned to this collection. You might also label these items by name for recording purposes.

  Before Session 1, select a few pairs of objects: some that are easily distinguished by weight and some that are more difficult, either because they are close in weight, or because their sizes are misleading (e.g., a big balloon and a small rock).

- Things to Balance: Gather about 20 lightweight items (half pound or less) that fit into your balance pans (e.g., pencil, crayon, box of paper clips, pad of stick-on notes, deck of cards, cassette tape). Note that the pan balance may not be sensitive enough for items that are very light, such as a single paper clip. Label each item with a colored sticker (a second color) and possibly by name.

- Prepare at least six balance stations, each with one pan balance, several objects from the Things to Balance collection, and 2–3 small containers of identical objects. You need not have exactly the same items at each balance. Post a piece of chart paper near the balance stations where students may record observations.

- After students have finished with the Things to Weigh by Hand collection in Session 4, fill two plastic grocery bags with several objects each (for demonstration). One bag should be heavier than the other.

- Duplicate the following student sheets, located at the end of this unit. If you have Student Activity Booklets, no copying is needed.

  Family Letter (p. 98): 1 per student. Sign and date before copying.

  Student Sheet 1, Which Is Heavier? (p. 99): 2 per pair

  Student Sheet 2, Something to Weigh (p. 100): 1 per student, homework

  Student Sheet 3, Heavier, Lighter, the Same (p. 101): 1 per student

  Student Sheet 4, Finding Things That Balance (p. 102): 1 per student, homework

- If you plan to provide folders in which students will save their work for the entire unit, prepare these for distribution.

Sessions 1 and 2

Exploring Weight

Materials

- Prepared balance stations (at least 6 per class)
- Things to Weigh by Hand (object collection)
- Pairs of objects selected for demonstration
- Student Sheet 1 (1 per pair)
- Posted chart paper
- *Who Sank the Boat?* or another book related to weight
- Student Sheet 2 (1 per student, homework)

What Happens

Students begin their exploration of weight by hefting objects in their two hands and using a pan balance for comparison. As a class they read a book related to weight and become familiar with the terms *heavier, lighter,* and *balanced.* Their work focuses on:

- hefting objects to compare their weights
- using a pan balance to compare weights
- developing language about weight

Activity

Introducing Which Is Heavier?

Most of the time during the first two sessions will be spent on Choice Time with two activity choices. To introduce the first activity, you will need the collection of Things to Weigh by Hand, including selected pairs of objects, and a copy of Student Sheet 1, Which Is Heavier?

For the next few weeks in math class, we're going to be measuring things and comparing them. We can find out lots of different things by measuring something. *[Hold up one of your objects.]* **For example, what could we find out about this [shoe] by measuring it?**

Encourage students to think of different types of measurements and ways of measuring. For example, we might find out *how long* the shoe is, *how wide* it is, *how far around* it is, *how much* sand it could *hold,* and how *heavy* it is. Some students may not be familiar with the term *measuring,* while others may think it is always "something you do with a ruler." See the **Teacher Note,** About Weight, Capacity, and Length (p. 13), for a discussion of how students will be measuring different characteristics of objects.

❖ **Tip for the Linguistically Diverse Classroom** As each type of measurement is suggested, demonstrate with gestures and actions the particular measure being discussed. For example, run a finger along the shoe from end to end to indicate *how long,* use thumb and forefinger to indicate *how wide,* and so forth.

For the next few days, we are going to be weighing and balancing things. What is weight? Where do you see weight being measured?

Students share their ideas about what weight is. They may talk about *heavy* things and *light* things. It is likely that they have seen food being weighed at the supermarket, or people being weighed at the doctor's office or on a bathroom scale. Some students may volunteer how much they weigh, although this can be a sensitive issue for some.

What's something heavier than you are? What's something lighter than you are?

Call on several students to name objects that they perceive as lighter or heavier than they are.

How can we tell how light or how heavy something is?

To find the weight of an object in standard units, such as "10 kilograms" or "8 ounces," we need a scale. However, we can get a general sense of how heavy or light it is by simply picking it up or comparing it to something else.

Let's look at a few pairs of objects and make some predictions about which one is heavier.

For the first comparison, choose one of the more obviously different pairs of objects you've selected. Write the names of those objects on the board in the format on the student sheet. For example:

```
Which is heavier?
  stapler_____   or   pencil_____
```

Students share their predictions, explaining why they think one object is heavier.

Sessions 1 and 2: Exploring Weight ■ **5**

Let's have someone come up and compare these two objects. You may want to close your eyes so you can concentrate on how heavy the objects feel.

Give the volunteer one object to hold in each hand. Then, while the volunteer explains which object feels heavier, pass the same two objects around the room so other students can get a feel for their weight. On the board, circle the object that most students think is heavier. Since the weight of an object isn't always apparent from its size and shape, students need to develop a physical sense of weight by hefting objects of different weights and comparing them in this way.

Repeat this process with another pair of objects that are more difficult to distinguish. Again, write the items being compared on the board.

```
Which is heavier?
(stapler)            or    pencil
cassette tape        or    chalk
```

As students make predictions, they might refer to the sizes of the objects:

   The cassette tape is heavier because it has stuff inside and it's bigger.

They might rely on their own previous experiences holding these objects:

   I've felt the chalk before and even with the tape inside the plastic case, the tape's still lighter.

Ask students how they can tell with their hands whether something is heavier or lighter. They may talk about a heavy thing "pulling your hand down more" or say that it's more tiring holding a heavier thing than holding a lighter thing.

Later today and tomorrow, you will have a chance to work with a partner to figure out which of two objects is heavier. Some pairs of objects will be harder to tell than others. You'll use this sheet [hold up a copy of Student Sheet 1] **to keep track of what you find out.**

Show students the collection of Things to Weigh by Hand, and explain how they can identify objects in this collection by the color of sticker on them.

You'll choose two objects from this collection. Then you'll write their names or draw the objects in the first two blanks on this sheet. After both of you have held the objects in your two hands, talk about which felt heavier. Then circle the one that you think is heavier, the way we just did in class. If you don't know or if you and your partner disagree, write a question mark after those two objects.

In the activity Using a Balance at the beginning of Session 4, students will use the balances to compare some of the object pairs that were difficult to distinguish just by hefting them.

Activity

Introducing the Balance

The second Choice Time activity is Exploring Balances. To introduce this activity, gather students around one of the balance stations, or bring those materials (a balance, several Things to Balance, and the containers of identical objects) to the meeting area.

Holding objects in our hands is one way to figure out if they are heavy or light. This balance is another way. Have any of you seen a balance like this one? Do you know how it works?

Encourage students to describe anything they know about balances. Where have students seen balances before? Have they used them? Can they identify the arm and the pans? Some students might mention a seesaw as a kind of balance.

Suppose I were to put this roll of tape in one balance pan. What do you expect will happen? Then what if I put a small rock in the other side? What would happen?

After students share their predictions, try this experiment, balancing a roll of tape against a small rock (or similar objects). Ask students to describe what happened and to explain what that tells them about the weights of the objects.

So Jacinta said the side with the rock went down more. Does that tell you anything about the weight of the tape and the weight of the rock?

If students do not mention the idea of the heavier side being lower when we are balancing things, remind them of their experiences hefting different objects, and the sensation of their hand being pulled down more when they hold a heavier object.

Today you will have a chance to explore how the balance works. You can use any of the objects at your balance station. *[Point out both the Things to Balance and the small identical objects.]* **What are some things you might try?**

Students may suggest trying to make the two sides balance, or putting in something heavy to see it "tip down." Or, they may simply want to see what happens with different objects.

Activity

Choice Time

If you and your students have been using the *Investigations* curriculum, your class will be familiar with the Choice Time format. During Choice Time, more than one activity is offered. Students move among activities, often returning to an activity more than once. You will need to help your students choose what they will go to first, so that you have a reasonable distribution of students between the two activities, and students will need to have guidelines for when and how they change activities. For more information about how to set up these sessions, see the **Teacher Note,** About Choice Time (p. 84).

Post a Choices list, explaining that these are the two activity choices that students may work on during the rest of this class and the next.

To record their work, students can draw and/or write about their discoveries. Students should use a recording method that makes sense to them and is fairly efficient for them. The majority of students' time should be spent on weighing and balancing rather than on recording.

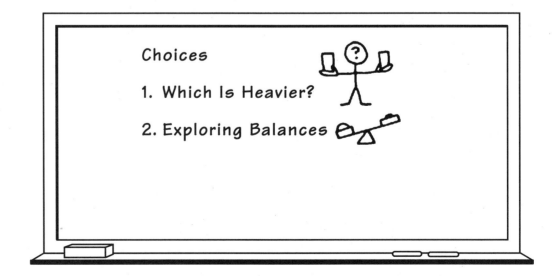

Investigation 1: Weighing and Balancing

Choice 1: Which Is Heavier?

Materials: Things to Weigh by Hand collection; copies of Student Sheet 1, Which Is Heavier? (1 per pair)

Students work in pairs, hefting pairs of objects. They record on the student sheet each object pair they try and circle which is heavier. If they do not know or if partners disagree, they write a question mark after that pair.

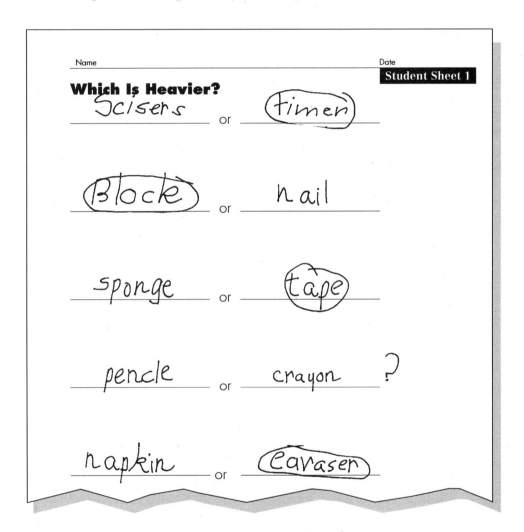

Choice 2: Exploring Balances

Materials: Balance stations (at least six, each with a balance, several Things to Balance, and containers of identical objects); posted chart paper

Students explore how the balances work, using the objects at each balance station. Help them record at least one question or discovery on a piece of chart paper posted nearby.

Sessions 1 and 2: Exploring Weight ■ 9

The more time students spend using the balances the better. Ideally, you would have enough balances for at least half of the class, working in pairs. If it is possible to buy or borrow more balances, then students have more opportunities to explore this device.

Observing the Students

As students are working, notice how they are using their own sense of what feels heavier and lighter.

Which Is Heavier?
- How are students predicting which objects are heavier or lighter? What information are they using to make their decision? Do they use what they know about what the objects are made of? the size of the objects? other information from previous experiences?
- Are students developing a feel for which items are heavier than others? Are their predictions realistic?
- How are students using language to talk about weight?

Exploring Balances
- How are students using the balance? What questions are they asking?
- Are students becoming comfortable with the idea that the "down" position on the balance indicates a heavier object than the "up" position? Do they know what the balance looks like when it is even or balanced?
- Are students predicting which items are heavier? Do they have a feel for which items are clearly heavier than others? What information are they using to make their decisions?

As you circulate through the classroom, notice what issues arise as students work with the balances. For example, do they notice that the balance may not be exactly level when there is nothing in it? If you have balances that can be adjusted, do they know how to adjust the empty balance? When they work with two identical objects, do they notice that it matters where in the pan the objects are placed, in order to balance them? Are they aware that the balance arm needs to stop swinging before they decide whether an object is heavier or lighter? Talk with students about their expectations as you circulate. List students' observations about these issues on the chart paper, and plan to raise some of the more common issues during the activity in Sessions 3–4, Introducing Balance Comparisons (p. 17).

Activity

Who Sank the Boat?

At the end of Session 2, gather students together and read aloud a book about weight. *Who Sank the Boat?* by Pamela Allen (Sandcastle, 1982) is ideal, but if you cannot get this book, read another book related to weight, such as *Math Counts: Weight* by Henry Pluckrose (Childrens Press, 1995) or *Just a Little Bit* by Ann Tompert (Houghton Mifflin, 1993).

In *Who Sank the Boat?* various farm animals climb into a small rowboat, one at a time, making the boat sink lower and lower in the water. At the same time, the animals *balance* each other's weights by getting into opposite ends of the boat. While you are reading and when you finish, ask students about what happened as each of the animals got into the boat. Refer back to the pictures.

What happened to the boat when the cow got in? Why did the front of the boat go up in the air? What does it mean when the book says that the donkey *balanced* the cow's weight?

What happened to the boat as more and more animals got in?

Students are likely to notice that at some times one side of the boat is up and the other side is down. At other times, the boat is balanced. If there is time, read the book a second time, asking students to pay attention to the change in the water level on the sides of the boat.

Is there anything about the story of the animals in the boat that reminds you of the balances?

Students may talk about the boat and the balance pans "sinking" or going down as you put more objects in them. They might also note that in both the boat and the balance, the lighter side rises up as the heavier end sinks lower.

Discuss any questions that students raise about the book, and consider using the balance to model their questions. For example, a student in one class wondered, "If the mouse had gotten in carefully, would the boat still have sunk?" Read the **Dialogue Box,** Who Sank the Boat? (p. 14), to see how one teacher handled this. Another class explored the question, "If the animals had gotten into the boat in a different order, would the boat still have sunk?"

Sessions 1 and 2 Follow-Up

Something to Weigh Send home the signed family letter or the *Investigations* at Home booklet to introduce your work on this unit. Also send home Student Sheet 2, Something to Weigh, which is a request for students to bring in an object from home that is small enough to fit in the pan of a balance. They will use these objects (along with lightweight objects from the classroom) during Sessions 3 and 4.

Weight Book If you can get the book *Math Counts: Weight* by Henry Pluckrose (Childrens Press, 1995) and did not use it for the in-class reading activity, read it with your group. Help them identify the different types of scales pictured, including the balances, and compare them to your class balances. Some of the questions and ideas in this book may be difficult for first graders. Focus on the pictures and descriptions of different situations in which weight is measured, and use these as a way for students to continue to share their own experiences with measuring weight.

About Weight, Capacity, and Length

Teacher Note

This unit introduces students to measuring and comparing the weights, capacities, and lengths of a variety of objects. Students' life experiences have already given them some ideas about these kinds of measurement. They feel how heavy and light things are; they notice how much of something it takes to fill containers; and they constantly experience a sense of length, width, and height as they look at the people and objects around them. During these investigations, help students link to the class activities the experiences and knowledge they already have.

Weight is not a characteristic that can be seen. In fact, judging weight by size and shape can be very misleading. Students need repeated experiences, hefting many objects, to develop their perception of what weight is so they can distinguish it from other characteristics. In the first investigation, Weighing and Balancing, students decide which things are heavier or lighter than others by using their individual perception of weight. They also learn to use an important tool, a pan balance.

Capacity is the amount of space inside a container (sometimes confused with *volume,* or the amount of space a solid takes up). In the second investigation, Filling, students explore capacity by filling containers with *discrete* (countable) objects such as cubes, as well as with *continuous* substances such as sand or water. To figure out which container holds more, they compare them directly, or count in nonstandard units (for example, cupfuls of sand) the amount needed to fill each one. They begin to see how specific numbers of units enable them to more clearly describe and compare capacities. Students use what they know about numbers, as well as size and shape, to predict and to justify their answers.

The length of something is a measure of that object from one end to the other. As a linear distance, length is more visible than either weight or capacity. In the third investigation, Measuring Length, students begin a series of activities that continues through *Investigations* grades 2 and 3: They compare lengths directly and also measure the lengths of objects in the classroom with nonstandard units (such as *cubes* or *footsteps*). In grade 3 they begin to measure with standard tools (rulers, metersticks), using standard units common to our society (inches and feet, centimeters and meters).

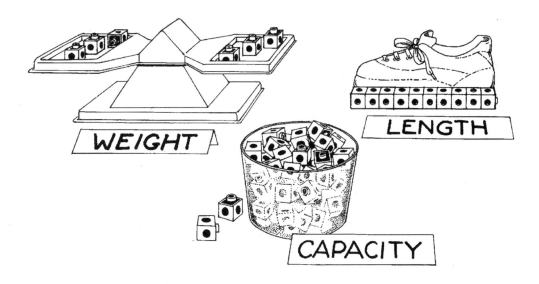

DIALOGUE BOX

Who Sank the Boat?

After hearing their teacher read the book *Who Sank the Boat?*, this class discusses a question that a student raises at the end of the story.

Donte: If the mouse had gotten on carefully, would the boat still sink?

[Some students say yes; others say no.]

Jonah: If it got on in the middle it would've been even.

So you think it depends *where* he got on, more than *how* he got on.

Garrett: I think it would sink, because a lot of them are on the boat.

Leah: Yeah. Because first the cow put weight on the boat, then the donkey was more weight, and they're still fat and still on the boat.

Kristi Ann: If the mouse got on carefully, it wouldn't have put as much weight on the boat.

Iris: The dock was too high up, so he had to jump.

[The teacher gets a balance and a tub of pattern blocks.]

Here's my boat *[the balance]* **and here's the mouse** *[a green pattern block]*. **I'll try putting it in two ways: throwing it, and placing it carefully.** *[First the teacher has the block "jump" into the left pan, which goes down lower than the right pan.]* **What happened?**

Nathan: The side you put the mouse in got heavier because there's something in there.

Now let's try Kristi Ann's suggestion. I'm going to put the mouse in very carefully. *[The teacher removes the block and then places it in the left side very carefully.]* **Did it make a difference?**

Yukiko: No. It still went down. *[There is lots of agreement.]*

Jacinta: Even if it goes creeping in, it's still heavy when it gets there.

Luis: If you jump, you'd make a big splash and the boat could go down.

Yukiko: But in the book the mouse jumped into the middle. And it still sank.

Why is that?

Leah: Because there are too many animals on the boat.

William: It's like Jacinta said, even if it was just a little bit of extra weight, it could make it heavy.

Jonah: The mouse just made it too tippy, because he was jumping.

Even though the balance doesn't exactly model the boat situation, the teacher used it to stimulate students' thinking about what makes a difference when you are considering the weight of an object. There are many complicated questions embedded in this discussion, involving both mathematics and physical motions. For example, we know that the mouse is the same weight whether it jumps into the boat or crawls in carefully. However, *where* in the boat it enters (the middle or an end) may determine whether the boat stays balanced. And in the context of a boat in water, jumping in with enough force could in fact rock the boat enough to make it capsize. Discussion of the issues allows students to further develop their ideas about weight and balance and to use appropriate language to describe these ideas.

Investigation 1: Weighing and Balancing

Sessions 3 and 4

Comparing Two Objects

What Happens

Students talk about their use of balances in Sessions 1 and 2 and consider any issues that have arisen. They identify pairs of objects that they found difficult to compare by hefting and consider how the balance helps them compare these weights. During Choice Time, students continue hefting objects to determine their comparative weights, checking their results with a balance or against classmates' perceptions. They also use a balance to find objects that weigh *more than, less than,* and *the same as* a particular object they have chosen. Each student draws a picture showing how the balance looks for one of these comparisons. Their work focuses on:

- comparing the weights of objects by hefting
- learning to use a balance
- comparing the weights of objects using a balance
- representing the results of weight comparisons

Note: If students have not brought in an object from home to use at the balances, they may look in their desk or elsewhere in the classroom to find something small and light enough.

Materials

- Student Sheet 1 (1 per pair)
- Student Sheet 3 (1 per student)
- Balance stations (at least 6)
- Things to Weigh by Hand (object collection)
- Lightweight objects brought from home

Activity

Using a Balance

When you were using your hands to test and compare the weight of things, were there pairs of objects that you and your partner were sure about? What about objects that you were *not* so sure about?

Call on a few students and find out what they discovered. As students name some of the object pairs that gave them difficulty, ask if any other students compared these, and what they found out.

Why do you suppose some objects are harder to tell apart than others?

Students may suggest that some objects are too close in weight or even the same weight, or are too light to easily gauge by just holding them.

Choose a pair of objects for which there is no consensus. They should be small enough to fit in a balance. Pass them around so that everyone has a chance to feel their weight and to say which object they think is heavier and why, or whether it's too hard to tell. In one class, for example, a student thought that a roll of tape felt heavier than a barrette "because the tape has more layers and the barrette only has one layer of flowers." Her partner thought the barrette felt heavier because it was metal and metal is heavy.

When you can't tell which one is heavier, that doesn't mean you're doing something wrong. Sometimes it is very difficult for anyone to tell which of two objects is heavier just by feeling them. What could we do to figure out some of the harder pairs? Who has an idea?

If no one mentions using the balance, suggest it yourself.

How might this balance help us?

Try out the pair of objects on the balance and ask students to watch what happens. They may see one side falling lower than the other, or that the sides are almost even (if the objects are close to equal in weight). Ask students what information this gives them about the two objects.

Explain that during Choice Time for the next two days, students will continue to test objects by hand for the activity Which Is Heavier? One balance will be available to them for trying out any questionable comparisons. If the objects in question are too big to fit in the balance, they may ask another pair of students to try lifting their objects as a check.

Issues of Using the Balance Based on the information you gathered in Sessions 1 and 2, address any questions about students' difficulties with the balance. Refer to the posted chart paper (with students' discoveries and questions) as needed.

What problems came up as you were using the balance?

Students may have found that some objects are difficult to keep on the balance, such as large objects, or objects that roll. They may have realized that the arms need to be clear of other objects so that they may rise and sink freely. Some may point out that you have to wait until the arm finishes rocking to see what happens.

Does this device balance when nothing is in it? How can we tell if it's balanced?

Students may notice that "you can use that arrow to see if it's the same. If it's in the middle, it's the same." Or they may notice that the arms are straight across, neither higher than the other. (If the demonstration balance is adjustable and is not level, adjust it accordingly, explaining that students may need to do this sometimes when they are working with the balance.)

Next place two identical objects (such as hexagonal pattern blocks) in the balance, one in each pan.

Does it matter where you place the objects within each pan?

Students may have noticed that even when you put the same thing on each side of the balance, one side may go up while the other goes down. See the **Dialogue Box,** Balancing and Placement (p. 22), for an example of how one class discussed this issue.

Activity

Introducing Balance Comparisons

In the new activity for the upcoming Choice Time, Balance Comparisons, student pairs select a single object (the one that they brought from home, or another from the classroom) and use it as a standard of comparison as they weigh it against objects from the Things to Balance collections. To introduce the activity, gather students at a balance station. Borrow an object that a student brought in from home or choose one yourself from the classroom—for example, a box of paper clips. Show students a copy of Student Sheet 3, Heavier, Lighter, the Same, and read it aloud as you explain what they will be doing.

You and your partner are going to choose *one* object, like this [box of paper clips], and put it in one side of the balance. Then you will be doing the three things listed on this sheet:

1. Find one thing heavier.

2. Find one thing lighter.

3. Find one thing that weighs the same.

So let's say Susanna and I are partners. We take our object—this box of paper clips—and put it on one side of the balance. At the top of this sheet, we identify what we have chosen, either with words or by drawing a picture. *[Demonstrate at the board.]*

First, we need to find something that is heavier than our box of paper clips. When we find it, how should the balance look?

Ask someone to draw a quick sketch at the board, showing how the balance should look. Then, as students suggest objects to try, demonstrate one or two of their ideas. Compare the results to their predictions.

Which was heavier? Was it a lot heavier? How far down did that side of the balance go?

Remind students that they will be doing the process three times, to find something heavier, something lighter, and something the same as their object.

When you have found something for all three cases, pick one of these to draw. Draw the balance carefully so that someone else can tell what is happening in your picture. Be sure I can tell what the two objects are, and what happened to the balance when you compared them.

Suggest that students move away from the balance stations for the drawing step, freeing the balance for other students to use.

Activity

Teacher Checkpoint

Choice Time

Post a list of Choices, explaining that these are the two activities that students may work on during Sessions 3 and 4. Students continue to record by either drawing pictures, writing words, or both.

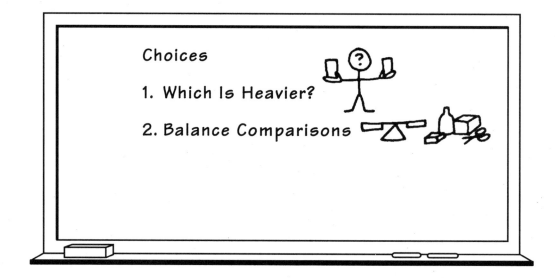

18 ■ *Investigation 1: Weighing and Balancing*

The Balance Comparisons activity is a Teacher Checkpoint, so make a point of observing all students at work on this second choice.

Students work on these activities until the end of Session 4. Expect that by the end of Choice Time, some pairs will have weighed more objects than others.

Choice 1: Which Is Heavier?

Materials: Things to Weigh by Hand; copies of Student Sheet 1, Which Is Heavier? (1 per pair); 1 balance available for checking (for the class to share)

Student pairs continue as in Sessions 1 and 2, hefting pairs of objects, recording each pair on a student sheet, and circling the heavier object. If they do not know which is heavier or if partners disagree, they write a question mark next to that pair and try them out on the balance when it is available to them. If the objects are too big for the balance, they may ask another pair of students to help them decide by hefting the two objects and telling which they think is heavier.

Choice 2: Balance Comparisons

Materials: Balance stations (at least 5); copies of Student Sheet 3, Heavier, Lighter, the Same (1 per student)

Students work in pairs but fill out their own student sheet. They choose an object (from home, their desks, or the classroom) and compare it with other Things to Balance at the balance stations. After finding one object that is heavier, one that is lighter, and one that is the same as theirs, they draw the balance showing one of these comparisons.

❖ **Tip for the Linguistically Diverse Classroom** As a visual reference, sketch three simple balances beside the problems on the student sheet. Beside the first problem, draw an arrow pointing to the side that is down (heavier); beside the second problem, an arrow to the side that is up (lighter); beside the third problem, draw a balance with both sides even.

Observing the Students

Although Balance Comparisons is a Teacher Checkpoint, try to observe students at work on both activities.

Which Is Heavier?
- How are students predicting which objects are heavier or lighter? What information are they using to make their decisions? Are they using their previous experiences? what they know about various materials objects are made of? the size of the objects?

- Are students developing a feel for which items are heavier than others? Are their predictions realistic?
- How are students using language about weight?

Balance Comparisons For this Teacher Checkpoint, observe students at work and look at their drawings to get a sense of their understanding of weight and how a balance indicates comparative weights.

- Do students understand how to use the balance? What strategies do they use to try to balance the scale?
- How are they reasoning about what will happen on the balance when one object is heavier?
- How do students record the results of balancing? Are their representations clear to others?

Note: Students may not be able to reliably predict which objects are heavier and lighter. It is age-appropriate for them to think that something larger must be heavier. Trial-and-error may be the predominant strategy for most students. As they compare the weights of a variety of objects, they are developing important ideas, experiences, and language about weighing and balancing. If you feel that students need more experience with balances, consider extending this work into science or other parts of your day and continue it for a longer period of time.

Sessions 3 and 4 Follow-Up

Balancing on a Seesaw If you can get the book *Just a Little Bit* by Ann Tompert and did not already read it for in-class work at the end of Session 2, read it aloud to the group. In this story, an elephant sits on one end of a seesaw, while a progression of animals joins a mouse at the other end to try to make the seesaw work. Discuss with students what is happening as each animal sits on the seesaw. You might ask them to predict what combination of animals it would take to balance an elephant.

Extension

DIALOGUE BOX
Balancing and Placement

While students are exploring the pan balances in Choice Time during Sessions 1 and 2, this teacher notices that some students are placing blocks very carefully in the balance pans, while others pay no attention to placement. To raise this issue, the teacher asks Eva to share what she's noticed about the placement of an object.

Eva: If you put the same things on the weighing scale, like, say, I have a block and a block of the same kind, the same weight, the same height, one will go up and one will go down.

The same exact blocks?

Eva: Yes.

[The teacher puts two identical blocks into the pans. The two sides look equal, and students comment that they are the same.]

Eva: But if you move them around, then they're not.

[Eva moves one block to the outside of the pan and the other to the inside. The side with the block on the outside goes down.]

Mia: This one's lighter because it's up there.

So does the other block become heavier when it's on the outside?

Garrett: It's the block on the inside. It's giving more weight to the other one.

[They put both objects near the outside, and they look nearly balanced.]

Which is more accurate?

Mia: Both on the outside?

Why, what if we put both on the inside? Would that be accurate?

Diego: Look at the arrow. If it's in the middle then it is.

What if we switched sides? What would happen? *[The teacher switches each block from one side to the other.]*

Garrett: They're the same because they're both on the inside.

So which block is heavier?

Students *[generally]:* They're the same.

I was watching some students who put their objects right in the middle of each pan. Next time you work with the balances, think about where you are putting your objects, and how that makes a difference.

While some students may expect two identical objects to balance, others may not. Issues about the effect of placement on balance, like on a seesaw, are quite complex. The teacher raised the issue so that students can build on their intuitive sense that identical objects weigh the same as they explore the effect of where they place objects in the balance pans. The teacher will continue to encourage students to experiment with placement as they work further with the balances.

Sessions 5 and 6

Balancing Groups of Objects

What Happens

Students work with partners on two Choice Time activities. In Balancing Grocery Bags, they use their developing sense of weight to balance groups of objects evenly in two bags. In Balancing Balances, they use the balances to make two groups of objects that weigh the same. Their work focuses on:

- using their sense of weight to balance a group of objects
- developing language about weighing and balancing
- adjusting the objects on both sides of the balance to achieve a balanced position

Materials

- Plastic grocery bags with handles (1 per student)
- Two bags holding preselected objects
- Things to Weigh by Hand (object collection)
- Balance stations
- Unlined paper
- Student Sheet 4 (1 per student, homework)

Activity

Introducing Balancing Grocery Bags

In this Choice Time activity, students will continue to explore weight by hefting and comparing different objects. Since weight can't necessarily be determined by looking, it's important that students have many experiences holding objects and feeling their relative weights.

To introduce the activity, hold the two grocery bags that you have filled with objects from the collection of Things to Weigh by Hand. Keep the rest of this collection available nearby.

Let's pretend that you are walking home from the grocery store, carrying two bags of groceries, one in each hand. Each bag has a few items in it. How can you get them to balance, so that you don't have one bag dragging on the ground with your arm hurting, while the other one is swinging in the air?

Students are likely to suggest moving objects from one bag to the other. Ask a volunteer to come up and hold the two grocery bags.

Pretend that your body is a balance. Feel how the weight of each bag is pressing against your hand. Which one feels heavier? How can you tell?

Ask another volunteer to come up and find a way to adjust the bags to make them balance. A student might take some objects out of the heavier side and put them in the lighter side, or switch objects from both sides. Ask the two students to explain the decisions they make. If necessary, they may add more objects to one side or the other, but caution students against working with too many objects at once. As students' arms get tired, it gets harder to tell which side weighs more.

How do the bags feel now?

Repeat the adjustment process using other volunteers. Once the bags feel about even to the student with the bags, ask another student to consider whether the bags feel about even.

This is one of the activities you'll be doing in Choice Time: You and a partner will work on balancing grocery bags that hold more than two objects. Once you agree that the bags are balanced, you'll record your solution on a sheet of paper.

Activity

Introducing Balancing Balances

To introduce the second Choice Time activity, gather students at a balance station or bring to the meeting area one balance, the containers of identical objects, and a few Things to Balance.

For the activity we call Balancing Balances, you will be trying to find a way to make the balance even. You can use any number of objects you want. Let's try it. I'm going to put the box of staples and three yellow hexagons on this side. What do you think I could put on the other side to make the balance even?

Take suggestions from the students about what objects to put in or take off the balance to get the two sides as even as possible. Point out that this is the same kind of thing they did with the grocery bags.

Let's find two more groups of objects that balance each other. You may adjust both sides of the balance. How shall we start?... Now what can we add or take away so the balance is even?

Ask volunteers to demonstrate their ideas. Some students might place a particular item on one side and try to balance by manipulating only the objects on the opposite side. Others might put a few items on each side and see what happens, then adjust by moving objects from one side to another, adjusting one side, or adjusting both sides. Any of these are good strategies.

Besides objects from our Things to Balance collections, you can use the identical objects in the containers at each balance station. For example, with this balance we have [washers, square tiles, and teddy bear counters]. If you want to, you can try balancing with just these small objects. For example, I figured out that [two bears] balance [one washer]. *[Demonstrate.]* How many bears would it take to balance two washers? four washers?

Encourage students to use what they know about numbers to figure out the problems you pose. See the **Dialogue Box,** Balancing Pattern Blocks (p. 30), to see how one group of students investigated quantitative comparisons.

As you work on Balancing Balances during Choice Time, look for different groups of objects that balance. You need to record at least one of your solutions to this problem on a blank sheet of paper. You may write the names of the objects or draw pictures to show the way your objects balance.

Activity

Choice Time

Post the Choices list and explain that these are the two activities that students may work on for the rest of this session and the next. Let students know that in both activities, it may be difficult to balance objects exactly, but they should be able to get pretty close, so that the weights are *about* even.

Near the end of Session 6, plan to announce the end of Choice Time and gather students for a whole-group discussion of their work (How Can We Tell If It's Balanced? p. 28).

```
Choices

1. Balancing Grocery Bags

2. Balancing Balances
```

Sessions 5 and 6: Balancing Groups of Objects ■ **25**

Choice 1: Balancing Grocery Bags

Materials: Plastic grocery bags (2 per pair); Things to Weigh by Hand collection; unlined paper (1 sheet per student)

Students work in pairs to balance small groups of objects in two grocery bags, one in each hand. They can move objects from bag to bag, add objects, or remove objects from the bags. Each student records one solution to this problem on unlined paper, using words or pictures. That is, they write or draw the objects they put in each bag to achieve a balance. Pairs may repeat the activity with different objects, but they record only one solution.

Choice 2: Balancing Balances

Materials: Balance stations; unlined paper

Student pairs find groups of objects that balance each other and record, using words or pictures, at least one of their groups of balanced objects. Students working at this choice will be solving one of three types of problems:

1. Balancing groups of objects from the Things to Balance collection. These students are asking the question, "How can I arrange objects to make the balance even?"

2. Balancing one object with a number of the identical objects (nonstandard units of weight). These students are asking questions like, "How many washers will it take to balance the tennis ball?"

3. Balancing two types of identical objects. These students are asking questions like these: "How many washers does it take to balance one yellow hexagon? If two washers balance one yellow hexagon, how many washers does it take to balance two yellow hexagons?"

Some students will need to focus primarily on the first type of problem, balancing groups of objects, exploring how to adjust both sides of the balance until they are even. Others will be ready to think about quantifying the measure of weight; for these students, pose questions that will lead them to explore *how many* of something balances something else, as described for the second and third type of problem.

Students can double-check their solutions by trading with another pair to try balancing each other's objects. During this activity, students can also continue to consider whether it matters *where* they place the objects in the balance pans.

Observing the Students

In these activities, students are adjusting groups of objects in the balance or in grocery bags and may be beginning to express weight as a quantity in nonstandard units. As you watch students work, you will be able to get a good sense of how they compare the weights of various objects and how they achieve a state of balance.

Choice 1: Balancing Grocery Bags
■ How are students adjusting their objects so that the bags balance? Are they using information about how heavy and light certain objects are in order to make adjustments to each bag? Are they using information from previous trials to adjust? When they try to make a bag lighter by taking out an object and it is then too light, what do they do? Can they think about how to adjust both bags in order to get them to balance?

Choice 2: Balancing Balances

- How are students approaching this activity? Do they have a plan? Are they using information from other weighings to figure out what to try next in order to balance their objects?

- Do students know how the balance looks when it is balanced? Do they know how it looks when one object is heavier and one is lighter?

- Do students use quantitative information to predict how much an object weighs? For example, if the box of paper clips was balanced by 20 yellow hexagons, do they have some idea that a pencil will be balanced by fewer hexagons?

- Are students using what they know about number relationships to adjust the balance? For example, if one hexagon weighs the same as two square tiles, how do they figure out how many hexagons it takes to balance six square tiles?

Note: It is age-appropriate for most students to use trial-and-error as a predominant strategy. Students are developing important ideas, experiences, and language about weighing and balancing as they participate in these activities. To give them more experience with balances, extend this work into science or other parts of your day.

While students are working, make a note of one particular selection of objects that students balanced in two grocery bags. For example, you may note that one pair balanced a tennis ball, a roll of tape, and two cubes in one bag with a small stapler and a box of paper clips in the other. Later, gather these same objects for use during the follow-up discussion. Also listen for questions, difficulties, and discoveries to discuss at the end of Choice Time in Session 6.

Activity

How Can We Tell If It's Balanced?

Near the end of Session 6, gather students for this whole-group discussion. Raise some of the questions that came up for students as they were working with the grocery bags. Take the group of balanced objects you gathered from one pair of students and mix them up, placing them into two grocery bags so that they do not balance. Ask a student to hold the bags in two hands.

This is a collection of objects that Shavonne and Kristi Ann figured out how to balance in their bags, but I've mixed them up so they don't balance. Let's see if you can balance them again.

Encourage students to come up and move objects around in the bags, without removing any or adding any new ones, until the bags are balanced. When they find a solution, tell the class whether the bags are balanced the same way that the original pair of students balanced them, or if this is a new way. If it is a new way, encourage students to find another solution.

Also take this time to discuss any other issues that arose during Choice Time. The **Dialogue Box,** Grocery Bag Balancing (p. 31), demonstrates issues that were discussed in one classroom.

Sessions 5 and 6 Follow-Up

Finding Things That Balance Send home Student Sheet 4, Finding Things That Balance. For homework, students try to make two bags that balance, using combinations of things in their kitchen (such as cans of soup, boxes of cereal). They draw or write about the combinations of objects they found that balanced.

How Many Washers? Students use the balance to further explore quantitative comparisons, using a nonstandard unit of weight such as washers or one kind of pattern block. For example, how many trapezoids does it take to balance each of five different objects (say, a pencil, a glue stick, a roll of tape, a wooden block, a pair of student scissors)? Can students tell from this information which of their five objects is the heaviest? the lightest?

What Balances Six Washers? Students create collections of objects that balance six washers. They group the collections in small plastic bags and place these in a large container; other students may try balancing each other's collections to double-check. This activity can be repeated using different numbers of washers.

DIALOGUE BOX

Balancing Pattern Blocks

This teacher introduces the Balancing Balances activity (p. 24) by raising some quantitative comparison questions as students explore *how many* of a particular pattern block shape will balance *how many* washers.

The other day, one pair of students discovered that three hexagons balance three washers. So my question is, how many hexagons would it take to balance *one* washer?

William: Two?

Susanna: No, two would be the same as two.

Tamika: One hexagon.

How do you know?

Tamika: Because three equals three, so maybe they weigh the same. *[The girl tries this, placing one hexagon and one washer carefully in the center of opposite pans. They appear to be balanced.]*

Given what you just saw, how many trapezoids would balance one washer?

Kaneisha: Two, because two of these blocks are one of the hexagon. *[The class tries this and they do balance.]*

What if we compared two trapezoids against a hexagon?

Susanna: It would be the same. *[Again, the class tries this and they balance.]*

This is a harder one. How many blues [large rhombus pattern blocks] would it take to balance two washers?

Diego: Five?

Chris: Six, because three on one. *[Chris and Diego each put three blues on top of a hexagon and see that the total is six.]*

This last question is really tricky. How many green triangles would it take to balance three washers?

[Fernando takes out some green triangles and places them on a hexagon.]

Fernando: It's 6, 12, ... 18, right?

How could we find out?

Fernando: It's 12 + 6. So *[counting on his fingers]* 13, 14, 15, 16, 17, 18. It's 18.

Why did you use those numbers?

Fernando: Because a washer is the same as a hexagon. And a hexagon has six triangles. So if you have three washers, then it's six and six and six.

[Fernando counts out 18 triangles and puts them on the scale opposite 3 washers. With some adjustment of where he places the blocks in the pan, he makes them balance.]

As these students work with the blocks and washers, they are building knowledge about how to quantify weight by establishing and using a unit of weight. They are also thinking about how the weight of one object gives them information about the total weight of several of the same object. This teacher is encouraging students to use their knowledge about number relationships in this new context.

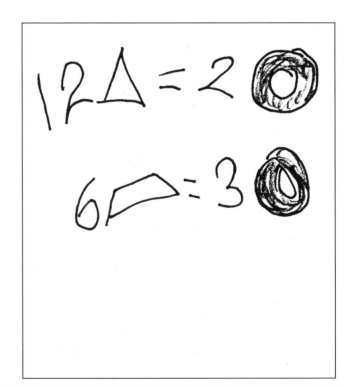

DIALOGUE BOX

Grocery Bag Balancing

This discussion took place after Choice Time in Session 6, when the students had been doing Balancing Grocery Bags in pairs. The teacher encourages students to explore what to look for when comparing the weights of objects in grocery bags. In one bag she has several objects: two small plastic cups, one washer, and an inflated balloon. In the other bag is simply one large Geoblock cube.

Which bag do you think is heavier? *[The teacher's right hand is lower than her left.]*

Kaneisha: Look at that bag [with the cube]; that's down more.

Can somebody help me balance them?

Donte: The bag that's heavier, put something in the other bag.

Tony: You can't take out of there [the bag with the Geoblock] because there'd be nothing left.

Tuan: Just go get something else. *[They add a box of paper clips to the bag with many things.]*

Now what happened? *[Now the teacher's hands are reversed, so the left one is lower.]*

Eva: That bag down there [with several things], that wasn't the heaviest, but now it is.

Jonah: Just take something out. [They take out the washer.]

How do you know when it's equal?

Eva: If your hands are equal.

Luis: It doesn't look equal. The block *[whose point seems to stick out of the bottom of the plastic grocery bag]* makes it heavier.

Donte: Of course this one [with the Geoblock] *looks* heavier because of that big block.

Susanna: The block's a little heavier because it's stretching down more.

Jamaar: The big block, it's a little too heavy, so it's kind of digging down into the bottom of the bag.

Mia: It goes down more, but your hands are equal.

If you're using bags in this experiment, is it important to look at the person's hands as you figure it out? Or is it more important to look at the bottom of the bags?

Several students: The hands.

Max: I think it's both. *[There is lots of agreement.]*

Andre: Maybe the bags are a little different.

Mia: But they are the exact same bag.

Shavonne: Maybe if you change the position of that balloon it would help. *[She moves it up so that the bag appears to have stretched taller. Most students think the bags now balance.]*

Kaneisha: The big block might have stretched out the bag a little.

Libby: Why don't we just feel it?

The two bags are passed around, and students agree that the weights feel about equal, even though the bag with the block looks lower because of the pointy shape sticking down.

INVESTIGATION 2
Filling

What Happens

Session 1: A Cupful of Sand Students experiment with how many spoonfuls of sand it takes to fill a paper or plastic cup. They discuss ways to make the filling process more consistent, repeat the experiment, and compare results.

Sessions 2, 3, and 4: Filling Space During Choice Time, students do two activities that involve filling two- and three-dimensional space. In Which Holds More Sand? they compare the capacity of two containers. In Block Puzzles, they explore ideas of area as they use pattern blocks to fill a shape outline exactly. Those ready for more challenge make their own Block Puzzles by identifying sets of pattern blocks that do and do not fit a particular outline.

Sessions 5, 6, and 7: Comparing Containers Choice Time continues for three more sessions as students continue working on Block Puzzles and two new activities that involve filling containers. In Comparing Bottles, students look for two bottles that hold the same amount of water. In Which Holds More Cubes? (a Teacher Checkpoint), students compare the number of cubes that different containers can hold. Students ready for more challenge can order the containers according to how many cubes each holds. The sessions conclude with a group experiment in which the class figures out which two of three containers hold the same amount of water.

Routines Refer to the section About Classroom Routines (pp. 87–94) for suggestions on integrating into the school day regular practice of mathematical skills in counting, exploring data, and understanding time and changes.

Mathemathical Emphasis

- Developing language to describe and compare capacity
- Comparing capacities
- Measuring and comparing capacity using nonstandard units
- Collecting and keeping track of data

INVESTIGATION 2

What to Plan Ahead of Time

Materials

- Plastic teaspoons: 1 per student and extras (Session 1)
- Small plastic cups (4–6 ounces): at least 2 per student (Sessions 1–3)
- Play sand in large plastic tubs: about 2 cupfuls per student (Sessions 1–4)
- Leveling tools (pencils, plastic knives, rulers): 1 per pair (Sessions 1–4)
- Newspaper to cover working surfaces (Sessions 1–4, optional)
- Pattern blocks: 1 bucket per 6–8 students; paper pattern blocks, pattern block stickers, or crayons, optional (Sessions 2–7)
- Interlocking cubes: class set, available in sets of 50–60 (Sessions 5–7)
- Wide-mouthed containers (e.g., coffee cans, small plastic buckets) holding 4–18 cups of sand: 2 per sand station (Sessions 2–4)
- See-through plastic bottles (e.g., soda, water, cooking oil, dishwashing liquid): 5 per water station, including two 1-liter bottles per station (Sessions 5–7)
- Empty containers (e.g., plastic cups, margarine tubs, soup cans, coffee cans, yogurt or cottage cheese tubs) holding 10–50 interlocking cubes: 20–24 for the class (Sessions 5–7)
- Water buckets, funnels: 1 of each per water station (Sessions 5–7)
- Food coloring (Sessions 5–7, optional)
- Three different-shaped containers, two having the same capacity (Session 7)
- Chart paper, unlined paper (available)

Other Preparation

- Before Session 2, set up two or three sand stations, each with a large tub of sand, two containers of different capacities, a plastic cup, and a leveling tool. Label containers with different symbols (e.g., star, heart). Tape down newspaper to protect surfaces.
- Before Session 5, set up two or three water stations, each with a bucket of water, a funnel, a plastic cup, two 1-liter plastic bottles of different shapes, and three other bottles. Adding a few drops of food coloring makes the water easier to see.
- Also before Session 5, label with letters the containers that hold 10–50 cubes.
- If you do not have manufactured paper pattern blocks or stickers, you can duplicate pages 112–117 on construction paper. Enlist adult help in cutting apart the shapes.
- Duplicate the following student sheets and teaching resources, located at the end of this unit. If you have Student Activity Booklets, no copying is needed.

For Sessions 2, 3, and 4

Student Sheet 5, Which Holds More? (p. 103): 2 per student (1 for Sessions 5–7)

Student Sheets 6–9, Block Puzzles A–D (p. 104): 1 per student

Student Sheets 10–11, Block Puzzles E–F (p. 108): 1 per student, optional

For Sessions 5, 6, and 7

Student Sheet 12, Comparing Bottles (p. 110): 1 per student

Student Sheet 13, Two Containers (p. 111): 1 per student, homework

Session 1

A Cupful of Sand

Materials

- Plastic teaspoons (1 per student and extras)
- Paper or plastic cups (1 per student and extras)
- Sand in tubs
- Newspaper to cover surfaces (optional)
- Pencils, rulers, knives for leveling
- Chart paper
- Unlined paper

What Happens

Students experiment with how many spoonfuls of sand it takes to fill a paper or plastic cup. They discuss ways to make the filling process more consistent, repeat the experiment, and compare results. Their work focuses on:

- estimating the number of units needed to fill a container
- experimenting with and describing techniques for filling
- collecting and interpreting data
- counting and keeping track

Activity

Filling Cups: Experiment 1

Distribute cups, spoons, and tubs of sand to students seated at tables or desks. Pass out unlined paper for recording.

Today we are going to do an experiment. Everyone has a cup and a spoon. All the cups are the same size, and all the spoons are the same size. Look carefully at your cup and spoon. See if you can estimate, or guess, how many spoonfuls of sand will fit in that cup.

Give students some time to think about this. Suggest that they compare the size of the cup and the spoon and try to visualize the number of spoonfuls in the cup.

Before we collect your estimates, you may each put two spoonfuls of sand in your cup. See if that helps you think about how many spoonfuls it will take to fill the cup completely.

Allow a moment for students to think or to talk with a neighbor about their guesses. Then record guesses on chart paper, in a column format, under the heading Estimates. Collect a good sample of estimates, but you need not call on every student.

What do you notice about these numbers?

Ask a few volunteers to comment on the estimates. They may notice recurring numbers that many students predicted, or numbers that seem particularly high or low. If students do not mention it, call attention to the range of numbers:

We have numbers all the way from 4 to 25, with a lot of numbers in the 20's.

If the issue of differences among spoonfuls arises ("He took more"), acknowledge this, but do not spend too much time on it yet; most students need to work through the experiment before recognizing the significance of different measuring techniques. If some students already realize that whether a spoonful is level or rounded is important, say something like this:

Keep thinking about your idea while we do the first experiment. Then we'll talk more about what you think a spoonful should be.

This is a good time to set up guidelines for the use of sand in the classroom. For example, when spooning sand into cups, students should be working over the newspaper or over the tubs.

Now you're going to work with a partner on our first experiment, to see how many spoonfuls it takes to fill a whole cup with sand. It's important to keep track of how many spoonfuls you put in. Every time you put in another spoonful, your partner will write it down in some way. You can use numbers, tally marks, or quick pictures. One person does the writing, while the other does the filling.

After you've filled the cup once, pour out the sand and do it again. This time, switch jobs: The person who did the filling now does the writing, and the person who did the writing now does the filling.

Students spend the next 10 minutes or so filling their cups with sand. As you observe students working, notice how they begin. Do they dump out the two spoonfuls? start from two? Notice how they are keeping track of their spoonfuls, and if they are paying attention to the size of their spoonfuls. Remind students to do this twice, so that each student in the pair has a chance to fill and to keep track of results.

Activity

What Is a Cupful of Sand?

When each pair has filled their cup twice, call the class together to share their data.

Now each of you tell me how many spoonfuls it took to fill your cup. I'll keep track of your results on this chart.

Call on each student and record the number of spoonfuls, in the order you receive them, under a heading such as *Experiment 1*.

What do you notice about these numbers?

Students will probably notice that the numbers differ even though everyone had the same size cups and spoons. They may also have something to say about how close together or far apart the numbers are. If students aren't sure what you are asking, you might ask some specific questions:

Are all the numbers the same? Are they pretty close or pretty far apart? What's the lowest number? What's the highest? Were you surprised when you saw how many spoonfuls it took to fill your cup? Was the number higher or lower than you expected?

Why do you suppose we have such different numbers?

Encourage students to describe different ways of measuring that they noticed or tried.

Note: Suggest that students share ideas by saying "Some people" or talking about what they themselves did, rather than using other students' names (such as, "Brady didn't even fill up his cup").

Some students may have used level spoonfuls while others used heaping spoonfuls. Some may have packed the sand down firmly in their cup, while others may not have filled their cup completely. If students suggest that the cups or spoons are different sizes, explain that if that were true, it would be a good reason, but these cups are all from the same package and are all the same size.

Part of the reason we got very different numbers is that we were measuring the sand in very different ways. We're going to try our experiment again. Let's talk about some ways of measuring that we all agree to, so that next time we will all measure the same way.

First, I'd like us to try using rounded spoonfuls because it seems easier for people to work with. *[Demonstrate what a heaping or rounded spoonful looks like.]*

Then, what should we call a "full" cup? How will you know when your cup is full?

Ask students for different suggestions to describe fullness. This can be an issue since students are so often told not to fill something to the very top because it will spill. In this case, they will need to fill the cup all the way up. They may say a cup is full "when the top is flat" or "when there's no room to put any more in." Ask students to demonstrate what they mean by showing their idea of a full cup of sand.

Sometimes people level off what they're measuring, so that they have exactly a cupful. Have you ever seen someone do that for cooking or baking?

Ask students for leveling suggestions, and then demonstrate some techniques for leveling off a cup of sand. For example, students might use the handle of the spoon, a plastic knife, a pencil, or a ruler to smooth the surface of the sand straight across the top of the cup. They can also shake the cup gently or tap it as they go along, to even out the mound of sand in the middle. Spilling too much sand out of the cup as they flatten it will make their count of spoonfuls less accurate. Keeping the sand level, especially as they reach the top, will help avoid this problem. Finally, decide on whether to pack the sand into the cup, and if so, how. Then discuss record-keeping techniques.

What were some ways that people found for keeping track of their spoonfuls?

Students share suggestions that include recording techniques as well as ways of working together. For example, one student suggested, "While you're scooping, you should be saying 1, 2...." See the **Teacher Note**, Working with a Partner (p. 39), for more information on difficulties that may arise as students work together to keep track.

Activity

Filling Cups: Experiment 2

Now that we've agreed on ways to fill the cup, let's try it again. We'll call this Experiment 2. Each of you will have a chance to fill the cup and a chance to record.

Students spend the next 10 minutes or so filling their cups with sand, using the agreed-upon measuring techniques. As in the first experiment, each pair fills their cup twice, taking turns as the filler and recorder.

Observing the Students

As you observe students working, notice the following:

- How are students keeping track of the number of spoonfuls? Are they working collaboratively with their partners?
- Are students using the agreed-upon techniques for measuring?

If students are having difficulty keeping track, note this, and suggest some alternative ways of working with a partner. If students are not using the agreed-upon techniques for filling, point out what you've noticed and encourage them to try again.

Students who have finished can begin cleaning up as other students complete their work. When everyone is ready, gather students together to share the data. Again record the number of spoonfuls on the chart under a heading such as *Experiment 2*. (Numbers from Experiment 1 should remain visible.)

What do you notice about these numbers? Does this list look different from our results in Experiment 1?

In spite of the standardization of techniques, there is still likely to be some variation in the results of this experiment. However, students may notice that the range of these data is smaller than the range of the data from the first experiment. It is likely that there will be a few very popular answers for the second experiment. If some students used level spoonfuls in Experiment 1, there may now be more numbers that are lower. Some students may point out that even though everyone used rounded spoonfuls, all rounded spoonfuls may not be exactly the same. See the **Dialogue Box**, Sand Experiments (p. 40), for an example of how this discussion might go.

If there is time left in the session, consider the Extension activity, which asks students what would happen if they used bigger spoons.

Session 1 Follow-Up

Extension

Big Spoon, Little Spoon Follow up the class experiments on filling cups with spoonfuls of sand by asking students what the numbers would look like if they used bigger spoons, and why. Some students may think that bigger spoons would result in bigger numbers, while others may reason that the bigger the spoon is, the fewer the number of spoonfuls that would fit. Invite pairs of students to repeat the Session 1 experiment using bigger spoons. List the results on chart paper, and ask students to compare the numbers of big and little spoonfuls needed.

Working with a Partner

Teacher Note

For the Filling Cups experiments in this investigation, figuring out with a partner how to keep track of the number of spoonfuls of sand in a cup is an interesting challenge for students. It may take pairs a few tries before they settle into a system that works. Even then, expect that there will be some inaccuracies in students' results.

You are likely to see a variety of recording methods as students work. While one student is scooping up sand with the spoon, the other student will be writing, using numbers, tallies, check marks, or other symbols. Difficulties can arise both for the filler and for the recorder. Fillers, for example, may not keep an eye on their spoonfuls as they're measuring. Are they consistently using heaping spoonfuls, or are they spilling a lot of the sand before it gets into the cup? If students are having difficulty keeping the sand on the spoon, encourage them to hold the cup over or next to the sand container and to follow the spoon with their eyes.

The coordination between the two students may be the hardest part of keeping track. Students are often so eager to dive into the sand that they both start filling, or the filler may start while the recorder is still writing their names on the recording sheet or doing something else. If this happens more than once, encourage students to talk with their partners before they start filling and make a plan. Who will go first? Will someone say the numbers aloud? Which partner? In attempts to coordinate their recordkeeping system, students will be trying to find a rhythm that works for both partners. They may find that the recorder tends to get ahead of the filler, writing at a (faster) pace that doesn't match the actual filling process. Or, the filler may get too far ahead of the recorder, perhaps because it takes longer to write bigger numbers than smaller ones, or because there is confusion over which number comes next. Agreeing on a plan, taking it slower, and checking in with partners are ways for students to deal with these difficulties.

During this investigation, students have several opportunities to work out these issues. While you will want to support students who are having difficulty, students also need to play an active role in figuring out a technique for filling, recording, and coordinating the process that works for them.

DIALOGUE BOX

Sand Experiments

This class has just finished Filling Cups: Experiment 2. Their results are on the board.

Experiment 1

22	22	19
19	19	21
21	23	27
22	25	23
18	18	19
22	23	25
21	21	24
29	22	20
24	20	

Experiment 2

16	19	17
18	15	16
17	17	18
17	20	19
16	19	17
17	22	17
14	14	18
17	19	16
14	18	

What do you notice about these two sets of numbers?

Libby: Some numbers are different and some are the same.

Kaneisha: Yeah, some are both *there* [in the Experiment 1 results] and over *there* [Experiment 2]. Like 22.

Max: Some are higher numbers and some are lower numbers.

Leah: I think 14 is the lowest … and 29 is the highest.

So why is it that all of you can use the exact same type of cup and the exact same type of spoon and get different numbers?

Nathan: Some people might take small scoopfuls and some people might take big scoopfuls.

How would that change their answer?

Luis: Bigger makes it more, no… less scoopfuls.

Hmm. Interesting. What else do people notice about the numbers?

Kaneisha: These [Experiment 2 results] all have 1's in them—like 16, 18, 19.

What does the 1 stand for?

Kaneisha: A teen. *[The teacher looks puzzled.]* A ten.

What do most of these numbers [Experiment 1 results] start with ?

Iris: Most of those are in the 20's.

So the numbers from Experiment 1 are higher. Why did it take more spoonfuls to fill the cup when we did Experiment 1?

Donte: Because it wasn't the same. We didn't have to do teeny or big heaping spoonfuls. Instead we did just any ones.

So why were they higher before?

Fernando: Because people put littler spoonfuls.

So if you put in littler spoonfuls, why do you get higher numbers?

Fernando: Because you put less amount in.

So it takes more…

Leah: Numbers.

What's a reason that it's important to do heaping spoonfuls, besides following directions?

Chanthou: Because if you get really small ones, you get a really big number, and it's hard to keep track, and you'll be the only one. And because when you do small ones, it doesn't really count, because they're really small.

After recording the data, teachers sometimes reorder both lists so that students can more easily see the differences. For example:

18 18 19 19 19 19 20 20 21 21 21
21 22 22 22 22 22 23 23 23 24 24
25 25 27 29

14 14 14 15 16 16 16 16 17 17 17
17 17 17 17 17 18 18 18 18 19 19
19 19 20 22

In this case, the teacher thought that the students were focusing on an important issue about how *smaller* spoonfuls result in more spoonfuls and decided not to interrupt the discussion to reorder the data at this time.

Sessions 2, 3, and 4

Filling Space

What Happens

During Choice Time, students do two activities that involve filling two- and three-dimensional space. In Which Holds More Sand? they compare the capacity of two containers. In Block Puzzles, they explore ideas of area as they use pattern blocks to fill a shape outline exactly. Those ready for more challenge make their own Block Puzzles by identifying sets of pattern blocks that do and do not fit a particular outline. Their work focuses on:

- developing and describing techniques for comparing the capacity of containers
- justifying why one container holds more than another
- describing capacities that can't be measured exactly in whole units
- filling a certain area with shapes
- noticing relationships among shapes
- justifying why a set of shapes does or doesn't fill an area exactly

Materials

- Pattern blocks (1 bucket per 6–8 students)
- Student Sheet 5 (1 per student)
- Student Sheets 6–9 (1 of each per student)
- Student Sheets 10–11 (1 of each per student, optional)
- Prepared sand stations
- Newspaper to cover surfaces (optional)
- Chart paper
- Paper pattern blocks and glue, or pattern block stickers, or crayons (optional)

Activity

Introducing Which Holds More Sand?

Gather students near one of the sand stations to introduce the first activity for Choice Time, Which Holds More Sand? Have a copy of Student Sheet 5, Which Holds More?

This activity choice is something like the one you did yesterday, filling the cups with spoonfuls of sand. This time, you will use plastic cups to fill the two containers you find at the station. You'll work with a partner to find out which container holds more sand.

What do you notice about the shapes of these different containers? Which do you think will hold more? Why?

As students describe the sizes and shapes of the containers, they may talk about roundness, fatness, length, and so forth. Encourage them to offer preliminary ideas about which container they think holds more and why.

How should we fill these containers? Should we use heaping cupfuls or level cupfuls?

Decide with students which technique the class will use. They may elect to take as big a cupful as they can, as they did with spoonfuls. Or they may choose to level off the sand at the top of the cup. If you have any concerns about how the activity went in Session 1 (such as problems with keeping track, or too much spilling), share them with students now.

Suppose you are almost done filling the container. You've put in 12 cupfuls, and it isn't quite full, but you can't fit in another whole cupful. Let's say it needs about this much more. *[Show about half a cupful of sand.]* **What do we call that?**

Students may describe this as "less than a cup" or "half a cup."

We've already counted 12 cups, and now we have this extra part of a cup, so what would we call the total?

Students may suggest "12 and a little bit," "between 12 and 13," or "12 and a half." If they do not bring up the idea of halves, introduce this terminology and the notation. While using fraction notation is not an emphasis at this age, it's appropriate for students to begin to see it used. Explain that this notation means 12 whole things and an extra half:

$$12\frac{1}{2}$$

See the **Teacher Note,** Talking About Halves (p. 48), for a discussion of introducing fractions to first graders. As these activities continue, encourage students to discuss what a half of something (like half a cup) looks like. Some students will be ready to consider how half is exactly between two whole quantities and how it is different from other "extra" parts that are less or more than a half.

Show the copy of Student Sheet 5 and explain how students will use it. Read through the sheet with them and demonstrate how to fill it out.

You can use the back of this sheet to keep track of your counting. When you know which container holds more, draw a picture of it. There is room on this sheet to test two different pairs of containers.

These recording sheets can help you review the work students are doing if you are not able to watch all of them as they work. If you find from their written work that some students have not figured out which container holds more, you may want them to repeat this activity while you watch and interact with them.

When you have finished at the sand station, make sure that the sand is back in the tub, and the cups and containers are empty.

Activity

Introducing Block Puzzles

Show students the pattern blocks and Student Sheets 6–9, Block Puzzles A–D.

During Choice Time, you will also have a chance to work on Block Puzzles. Pick an outline that you want to fill. Below each outline, you'll see two different charts. Each chart shows a different set of blocks. One set will fill in the outline exactly. The other set won't work. Choose one chart and take out all the pattern blocks that are listed there.

Show students one chart and read it with them, setting aside the right number of each type of block.

When you have all the blocks listed in that chart, double-check them. *[Demonstrate.]* **Try to use just those blocks to fill the puzzle shape.**

Remind students that they may have to try moving the blocks around in different ways to see if they fit. Be sure everyone understands the same thing by the term *fit:* The blocks need to fit exactly inside the lines of the outline, filling up the space entirely and not hanging over the edge.

When you finish trying one set of blocks, then try the other set, listed in the other chart. Even if you think you've found the set that fits, try the other set, to double-check. After you've tried both sets, fill in the blank with the number of the set of blocks that works. *[Show them where on the sheet.]* **You may work alone or with a partner.**

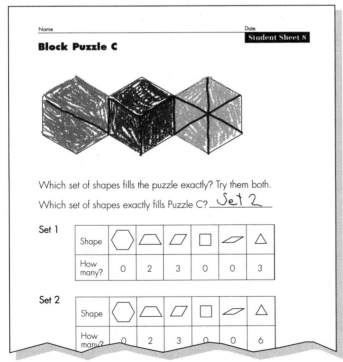

Sessions 2, 3, and 4: Filling Space ■ 43

Activity

Choice Time

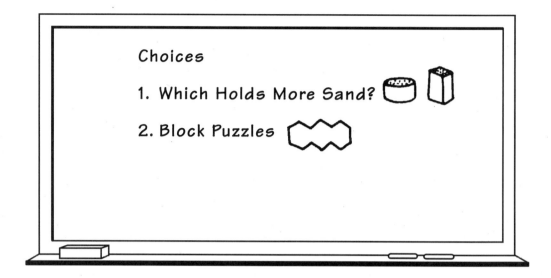

Post the two choices and remind students of their options. Explain that they will be working on these activities for the rest of this session and the next two sessions. Because fewer students can work at the sand stations at the same time, you will need to help students rotate after a reasonable amount of time so that everyone gets to this choice at some point in these three sessions.

Near the end of Choice Time, gather the class for a brief discussion of their work on Which Holds More Sand? (see Discussing Pairs of Containers, p. 46).

Choice 1: Which Holds More Sand?

Materials: At each sand station: newspaper to cover surfaces, a large tub or bucket of sand, two plastic containers that hold different amounts of sand (labeled with picture symbols), plastic cups, an object for leveling, and Student Sheet 5, Which Holds More? (1 per student)

One pair of students can work at each station. The students will need to figure out how to share the work. This will vary, depending on their approach to solving the problem. If they decide to fill both containers with sand and count the number of cupfuls needed, one person might record while the other fills the first container. Then they can switch roles as they fill their second container. However, if they take another approach and fill one container with sand, then pour that sand into the second container to compare, they will need to think about other ways to share the responsibility. For example, they might alternate pouring cups of sand into the container they fill. Whichever method they use, students record their results on Student Sheet 5.

Choice 2: Block Puzzles

Materials: Pattern blocks (1 bucket per 6–8 students); Student Sheets 6–9, Block Puzzles A–D (1 per student); paper pattern blocks and glue, or pattern block stickers, or crayons (optional). Also have available copies of Student Sheets 10 and 11, Block Puzzles E and F, for students who are ready for more challenge.

Some students may work individually if there are enough materials; otherwise, student pairs work on filling the outlines with the pattern blocks listed on each puzzle sheet and determine which set fits. They write the number of the set that worked. If necessary, remind students that blocks can't be piled on top of each other or stood up on edge. Optionally, they may use paper pattern blocks and glue, pattern block stickers, or crayons to record their solution to one or all of the puzzles they complete.

For More Challenge If you notice that some students are solving the Block Puzzles confidently, give them copies of Student Sheets 10 and 11, Block Puzzles E and F. On these sheets, students make their own Block Puzzle for other students to solve. They find a set of blocks that fits the outline and record the numbers of blocks in one of the empty charts. In the other chart, they record a set of blocks that does not fit in the outline. You may want to discuss with these students how to make a set that does not work but is close enough to make a good puzzle; students may have ideas based on their experiences with the other puzzle sheets. As students complete their own Block Puzzles, they can exchange with other students to solve.

Observing the Students

These two activities focus on filling space—both 3-D space (capacity) and 2-D space (area). Watch students work to get a sense of how they are building ideas about capacity and area.

Which Holds More Sand?
- How do students decide when the container is full?
- Are students keeping track, in a way that makes sense to them, of how many cups of sand they use? Are they able to count accurately? Do they double-check their counts?
- How do students compare the sizes of the containers? Are they counting cupfuls of sand? pouring sand from one container to another?
- How do students handle the situation when they need only part of a cup to finish filling the container? What do they record?
- Do students make reasonable arguments about why one container holds more?
- Do students know which numbers are larger than others or do they need a tool such as a 100 chart to help them decide?

Block Puzzles

- How are students filling the outline shape? Are they fitting blocks inside the shape exactly? Are they able to judge when blocks are hanging over the edge or when the space is not filled?
- When one arrangement of a particular set of blocks doesn't work, do students assume that set doesn't fit, or do they experiment with different arrangements? Do they brush off blocks after filling a shape and then randomly begin again, or do they try to make revisions in the design they have? How do they decide that a set definitely does not fit?
- Do students use relationships among shapes to reason about why one set fills an area and another set is too big or too small?

For students who need more practice, encourage them to repeat the activity, selecting different Block Puzzles.

For students who have gone on to work with Block Puzzles E and F:

- Do students easily find shapes to fill in the outline? Are they fluent in finding ways to make shapes fit together in the interior of the outline where it is not so obvious which shapes to choose?
- Do you see evidence that students know how to make the same shape in different ways? For example, do students see how to fill a trapezoidal shape with three triangles or with a blue rhombus and a triangle?
- How do students record their designs? If they are using paper pattern blocks, can they match the paper shapes to the blocks in the same arrangement, or do they start all over again with the paper shapes and make a new design?
- How do students make their own Block Puzzles? In particular, how do they figure out a group of blocks that does *not* work? Do they make a shape right on the outline that is either too large or too small? Do they use information about the collection of blocks that *does* work to make a set of blocks that *doesn't* fill the same area?

Activity

Discussing Pairs of Containers

At the end of this Choice Time, hold a brief class discussion about the pairs of containers from the Which Holds More Sand? activity. Hold up one pair of containers and ask:

How did you figure out which of these containers held more? Who can tell about the method they used? Who used a different method?

Students may need to use the containers of sand to demonstrate their ideas. If no one brings up the method of pouring sand directly from one container to the other, demonstrate this as another way to see which container holds more.

Were you surprised that this [star] container held more than that [heart] container? Why or why not? Can you tell this one holds more just by looking at it?

How did you keep track of how much sand each container held?

Who has something to say about what was hard about this activity?

If there is time, continue the discussion by looking at another pair of containers.

Sessions 2, 3, and 4 Follow-Up

Extensions

Capacity Book Read *Math Counts: Capacity* by Henry Pluckrose (Childrens Press, 1995) with your class. Help them identify the wide variety of containers pictured, including an egg cup, a thimble, a watering can, a perfume bottle, and a film canister. Ask students to estimate which container on a page holds the most, and which might hold the least. Focus on the idea of capacity as "the most a container can hold" rather than talking about particular units used to measure capacity. You might follow up with questions like these:

Which has the bigger capacity: a bathtub or a sink? the trunk of a car or the inside of a bus? the lake in the park or the swimming pool at the community center?

Creating Block Puzzles Students can create their own solutions for Block Puzzles A–D. They may record the number of blocks in their new solutions in any way they like, perhaps creating a chart like the ones below each puzzle. They may also record their solutions using paper pattern blocks or pattern block stickers, or by tracing the outlines of the blocks and coloring.

Teacher Note: Talking About Halves

The idea of halves is likely to arise quite naturally during work on this unit. In this second investigation, students will encounter situations in which spoonfuls or cupfuls of sand do not fit exactly into the container they are filling. This may lead to a discussion about which of two numbers (e.g., 16 cups or 17 cups) is closer to the amount a container can hold, and how we can tell. Often students have a tendency to choose the smaller of the two numbers so as not to go beyond the capacity of the container being measured, even if the amount is actually closer to the larger of the two numbers.

As situations like this one arise, ask students to think about ways we can describe an amount between two numbers. For example, we can say "about 17," "more than 16 cupfuls," "between 16 and 17," "16 and a little bit," and so forth. If the extra unit is in fact about a half, you can introduce the language and notation for one half (½). Explain, for example, that 16 full cups and a half cup is written this way:

$$16\frac{1}{2}$$

This notation will be unfamiliar to many of your students, and if they use it on their own, they may write it incorrectly. For example, many young students write 17½ to mean a half cup *less* than 17. Other students might reverse the numbers in the fraction and write 16 2/1. While first grade students are not expected to learn fraction notation, they can certainly begin to see it written and read correctly and begin to try it out themselves.

Halves may also come up during your work on Investigation 3, Measuring Length. When students are measuring lengths by sequentially repeating baby steps, hands, cubes, and other nonstandard units along an object, the measurement won't always "come out even." Again, students can explore ideas about halves. Some students will begin to distinguish between an "extra" amount that is close to a half—that is, exactly between the two whole quantities—and extra amounts that are much less or much more than a half. As second graders in the *Investigations* curriculum, students will spend more time on ideas about halves (and other fractions) in the unit *Shapes, Halves, and Symmetry*. For first graders, it is enough to recognize situations when one unit does not fit exactly into another, and to have a way of talking about this.

Pattern Block Shapes

Teacher Note

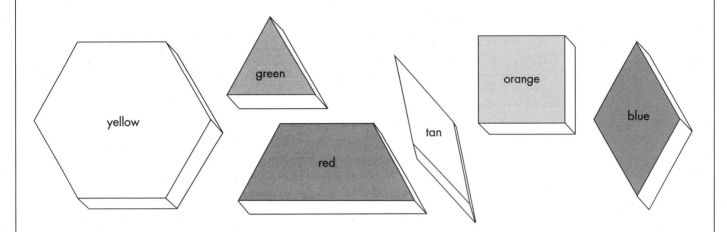

The pattern block set is made up of six shapes: a hexagon, a trapezoid, a square, a triangle, and two rhombuses. In most sets each shape comes in one color: the hexagons are yellow; the trapezoids are red; the squares are orange; the triangles are green; the narrower rhombuses are tan; and the wider rhombuses are blue. Since all pattern blocks of the same color are the same shape, it is very natural for students to identify them by color. In fact, teachers and students often name the blocks by using both color and shape, for example, "the yellow hexagon," "the orange square."

Many first grade students easily identify the green block as a triangle and the orange block as a square. The terms *trapezoid* and *hexagon* may be new to many first graders. They will learn to use the terms readily if you use them consistently yourself and help them remember the words when they forget. However, it is fine for students to use other informal language of their own. Young children typically refer to the blue and tan blocks as diamonds. This is fine, and there's no need to stop students from using this familiar term. As long as they are communicating effectively, let them use the language they are comfortable with while you continue to model the formal terms. These shapes are both parallelograms and rhombuses. The pattern block convention is to call them "the tan rhombus" and "the blue rhombus," because that identifies these shapes more precisely, focusing on their four equal sides. Following this convention, we identify these blocks as rhombuses in this unit.

Notice that in this unit, we speak of the pattern block shapes as if they were two-dimensional, even though they actually have three dimensions. The piece we call a square is not actually a square; it is a rectangular prism. Since the pattern blocks are all the same thickness, the activities focus attention on only one face of the block—the triangle, square, rhombus, trapezoid, or hexagon—and so the convention is to name the blocks by these faces, as most convenient and sensible for the way they are used.

Sessions 2, 3, and 4: Filling Space ■ 49

Sessions 5, 6, and 7

Comparing Containers

Materials

- Prepared water stations
- Interlocking cubes (class set, available in tubs of 50–60)
- Containers labeled with letters (for cube activity)
- Student Sheet 5 (1 per student)
- Student Sheet 12 (1 per student)
- Pattern Blocks and Block Puzzles from Sessions 2–4
- Three containers of different shapes, two with the same capacity
- Student Sheet 13 (1 per student, homework)

What Happens

Choice Time continues for three more sessions as students continue working on Block Puzzles and two new activities that involve filling containers. In Comparing Bottles, students look for two bottles that hold the same amount of water. In Which Holds More Cubes? (a Teacher Checkpoint), students compare the number of cubes that different containers can hold. Students ready for more challenge can order the containers according to how many cubes each holds. The sessions conclude with a group experiment in which the class figures out which two of three containers hold the same amount of water. Students' work focuses on:

- exploring measuring techniques for filling
- using a unit to measure capacity
- comparing capacities of two containers by filling with continuous substances (water) or discrete objects (cubes)
- relating size and shape to capacity
- comparing the capacities of more than two containers
- describing capacities that can't be measured exactly in whole units
- counting quantities to 50

Activity

Introducing Comparing Bottles

Explain that for Choice Time, one of the choices will be Comparing Bottles, at the water stations. Gather students at one of the water stations to introduce the activity.

There are five bottles at each water station. Your challenge is to figure out which two bottles hold the same amount of water. Can you make any predictions by looking at them?

Help students focus on the shapes and sizes of the bottles and discuss how they think the *tallness* or *width* of a bottle will affect how much water it holds.

What could you do to find out which two bottles hold the same amount?

As in their work at the sand stations, students may propose at least two methods for doing this comparison. Some may fill one bottle and then pour that water into another, seeing which holds more; others may fill a plastic cup with water and count the number of cups it takes to fill each bottle, comparing the totals.

Demonstrate at least one method. For example, first ask a student to pour some water from one bottle to another. (Choose two bottles of different sizes so that the class doesn't discover at this time the two that match.) Show how to use a funnel, explaining how this helps people be more careful about getting all the water into the bottle. When they pour water through a funnel, they will need to watch carefully the bottle they are pouring into so they can see when (if) it is going to overflow.

What does this experiment tell you about which bottle is bigger? How can you tell?

Students may say, for example, that the second container isn't all the way filled with water, so it has more space, so it's bigger. Or they may notice that all the water from the first container wouldn't fit in the second one, so the first one must be bigger.

Now choose a small bottle that you have already determined can't be filled with a whole (even) number of cups.

Some of you might decide to find out how many cups of water each container holds. Let's try that, and count how many cups of water this bottle holds.

Using a funnel, pour water into the bottle one cup at a time. Enlist student help in keeping track of the count. At the point when you finish filling the container with a partial cupful, ask students how you should count this partial cup. Is it just a little bit? Is it almost a full cup? Is it close to half a cup?

Use this demonstration to remind students about pouring carefully and holding the containers over the bucket. Careful pouring is important both for practical reasons and for mathematical ones: If too much water is spilled, students may not be accurately comparing the amounts of water in the bottles.

Show a copy of Student Sheet 12, Comparing Bottles.

You and your partner can use this sheet to keep track of what you're doing as you are working. When you think you've found out which two bottles hold the same amount of water, draw the shapes of the bottles and write their letters.

Activity

Introducing Which Holds More Cubes?

Note: This activity is similar to Which Holds More Sand? from the previous Choice Time. However, in this activity, students are filling the containers with discrete, countable objects rather than with a continuous quantity (like sand or water) that can be measured. Nonetheless, in both cases they are thinking about how much space there is in each container, how size and shape affect how much a container can hold, and how one size and shape compares with another (for example, can a tall, skinny container hold as much as a fatter, shorter one?).

To demonstrate the second Choice Time option, Which Holds More Cubes? show students two containers you have chosen and a supply of interlocking cubes.

If I filled each of these containers with cubes, which one do you think would hold more?

Ask students to explain their reasons for their choice. They may focus on the height of the container, the shape of the container, or other characteristics. Then fill one of the containers with individual (loose) cubes.

When you filled a container with sand, you carefully leveled off the top to make it exactly "full." Can we do that with cubes? How should we decide when the container of cubes is full?

Students share their ideas about what "full" means in this situation. They may talk about taking off the "extra" cubes and making sure the corners aren't empty, so the top is as flat as possible. They may suggest shaking the containers so that they don't leave big empty spaces.

Fill one of the containers with cubes. Then ask students to help you count as you remove the cubes, one at a time, and snap them together. Repeat for the other container. Hold up a copy of Student Sheet 5, Which Holds More?

This is the sheet you can use to keep track of your counts. You used this same sheet before when you were working with sand. What are some ways you could keep track?

Students might suggest using tallies, numbers, making cube towers of 10, counting by 1's or 2's, and so forth. To show which container holds more, they may draw a picture of it.

Explore with students how they might answer the question on the student sheet, "How did you figure it out?" For example, they could draw a picture of both containers and write the number of cubes on each one. Or they might draw the two towers of cubes that filled the two containers. Some may want to write a sentence, like one of these:

B had 24 and E had 18, so B had more.

B had 24 and they couldn't all fit in E, so B is more.

Activity

Teacher Checkpoint
Choice Time

Post a list of the activity choices. Choices 1 and 2 are the activities just introduced; for Choice 3, students continue their work on the Block Puzzles as begun in the previous Choice Time. Briefly remind students of their options and explain that they will be working on these choices for the rest of this session and the next two sessions.

Because fewer students can work at the water stations at the same time, help students rotate after a reasonable amount of time so that everyone gets to Choice 1 at some point in these three sessions. Help students choose where they will start.

Which Holds More Cubes? is a Teacher Checkpoint, so try to observe all students working on this second choice.

Plan to gather the class together at the end of this Choice Time for a discussion of their work on Comparing Bottles.

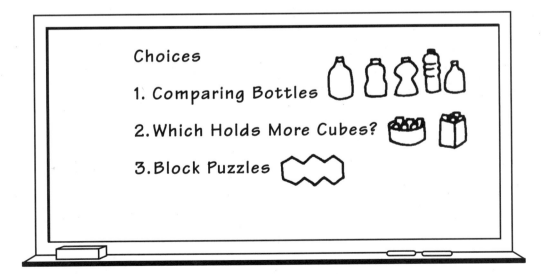

Choice 1: Comparing Bottles

Materials: At each water station: buckets of water, 5 plastic bottles, funnels, plastic cups; Student Sheet 12, Comparing Bottles (1 per student)

Student pairs compare how much the five bottles will hold, either by pouring water from one to another or by using a third container (such as a cup or another bottle) as a nonstandard measuring unit. They decide which two bottles hold the same amount, record the letters of those bottles on their recording sheet, and draw the two bottles.

❖ **Tip for the Linguistically Diverse Classroom** As students are talking to one another during this activity, model how English-proficient students can make their discussion comprehensible to second-language learners. For example, as you watch a student pouring water into a bottle, say:

The water *[point]* **fills this bottle** *[point to entire bottle]* **to here** *[point to water level]*.

The recording sheets for this activity can help you review the work students are doing if you are not able to watch all of them as they work. If students haven't found the two bottles that match, you might ask them to repeat this activity while you watch and interact with them.

Choice 2: Which Holds More Cubes?

Materials: A variety of containers labeled with letters; interlocking cubes, available in tubs of 50–60; Student Sheet 5, Which Holds More? (1 per student). Also have available unlined paper for students who are ready for more challenge.

Show students where you can find the containers. Working individually or in pairs, students choose two containers at a time to work with. They write the corresponding letters on Student Sheet 5 to identify the ones they chose, and then fill each container with cubes. It may help some students keep track when they are counting and comparing if they use a different color to fill each container (such as all red cubes in Container A, and all blue in Container B).

Students count and compare the cubes to see which container holds more and record their findings on Student Sheet 5. Encourage students to include comments about the result, such as, "I knew B would hold more because it's taller and bigger around."

Encourage students to do this activity several times, using different containers. Remember that this activity is a Teacher Checkpoint.

For More Challenge If some students are solving the Which Holds More Cubes? problems confidently, suggest that they choose four containers and put them in order, from the one that holds the least cubes to the one that holds the most (or the reverse). To help them order the numbers of cubes, they may refer to 100 charts or number lines you may have in the classroom. To record their work, they draw the four containers in order on unlined paper, labeling each drawing with the corresponding letters and with the number of cubes that container holds.

Choice 3: Block Puzzles

Materials: Pattern blocks; Student Sheets 6–11, Block Puzzles A–F (copies remaining from previous Choice Time); paper pattern blocks and glue, or pattern block stickers, or crayons (optional)

Students continue their work on any Block Puzzles they have not yet completed. Optionally, they may use paper pattern blocks, pattern block stickers, or crayons to record their solutions. Remember that Block Puzzles E and F are designed for more challenge. On these sheets, students make their own Block Puzzle for other students to solve.

Observing the Students

See p. 46 to review what to look for while students are working on Block Puzzles. Look for the following as you watch students filling containers in the other two activities during this Choice Time:

Comparing Bottles
- What strategies are students using to compare how much the bottles hold? Are they using a cup, the bottles themselves, a third bottle? Can they explain their reasoning? For example, when they pour all of the water from one bottle into another and there is room left in the second bottle, do they know which of the two holds more? Do they double-check?

- How do students keep track of which comparisons they have made? Do they keep trying pairs of bottles randomly, or do they use information from one comparison to help them with the next comparison? How do they decide which comparisons to make next?
- Are students consistently using full cupfuls of water? If they use less, are they describing this as a *half* a cupful?

If some students are having difficulty comparing all five bottles, provide a hint by showing them one 1-liter container and asking them to find which one of the remaining four bottles holds the same amount as that one. Or, consider having students compare only three bottles at a time.

Which Holds More Cubes? As you observe students at work on this Teacher Checkpoint activity, you will be able to get a good sense of how they are exploring the ways size and shape are related to capacity.

- Can students talk about why one container might hold more in terms of the size and shape of the container?
- Do students think about what it means for a container to be "full" of cubes? Do they try to make both containers they are comparing "full" in the same way?
- Do students connect the cubes or use another method to make sure they have counted each cube exactly once? Do they count cubes accurately?
- As students repeat the activity, do they begin to develop a better sense of how many cubes a particular container might hold and how two containers might compare?

If some students would be better off working with small numbers, provide smaller containers for them to compare.

Looking carefully at the written work on Student Sheet 5, Which Holds More? may give you additional information on the strategies students are using to compare containers.

Activity

Discussing Our Work with Bottles of Water

For this discussion at the end of Choice Time, you will need a variety of bottles from the water stations and a bucket of water. Gather the class where everyone can see these and where you and a few students can work with them. Also have available the three containers you have kept aside for this follow-up discussion.

Over the past few days, you've been using water and cubes to fill up different containers. Let's talk about some of the things you discovered as you were working with the bottles of water.

When you were filling these bottles with water, how did you measure which held more? How did you figure out which two bottles held the same amount?

Students will be interested in having confirmed which two bottles at each station hold the same amount of water. After one or two volunteers share their conclusion, ask these students to demonstrate, using bottles and water, *how* they know those two bottles hold the same amount.

Can you tell by looking which bottles held the same amount? Why or why not? Can you tell by looking that some of these bottles for sure don't hold the same amount? Which ones? How do you know they couldn't possibly hold the same amount?

Hold up bottles that students indicate. Encourage them to talk about the shapes and dimensions of the bottles and how they think those characteristics might affect how much the bottle can hold.

❖ **Tip for the Linguistically Diverse Classroom** Encourage all students to indicate with their hands the shapes and dimensions they are talking about to help make this discussion comprehensible.

Now show students the three new containers they haven't worked with. Ask them which two of these they think hold the same amount. After they have made their predictions and given their reasons for which two they think are the same, ask students to suggest a method for determining which two hold the same amount. Together, carry out at least one method they suggest. The **Dialogue Box,** Comparing Cupfuls (p. 58), demonstrates how this experiment went in one first grade classroom.

Sessions 5, 6, and 7 Follow-Up

Two Containers After Session 5, send home Student Sheet 13, Two Containers. Students find two containers at home. They figure out which holds more by filling them with water (or another substance that is easy to pour). Some students might want to use a measuring cup, a small paper cup, or some other small container to fill the containers and keep track of how many of the smaller container was needed to fill each larger container. Students record their results by drawing the two containers and writing what they did and what they found out.

DIALOGUE BOX

Comparing Cupfuls

At the end of the final Choice Time in Investigation 2, this teacher is leading students in the follow-up discussion. As a group, the students are trying to figure out which two of three new containers hold the same amount of water. The containers are (1) a wide flowered glass with straight sides, (2) a green mug with sloping sides, and (3) a red plastic drinking glass, taller than the other two, with sloping sides. Students are sitting in a circle around a tub of water with the three containers at hand.

Look carefully at these three containers. Which two do you think hold the same?

Yanni: I think the flowered one and the red one are the same. The green one holds more.

Why do you think that?

Yanni: The green one looks taller and wider at the top.

Tamika: Not the flowered one.

Why not?

Tamika: It's bigger, it's wider.

Chanthou: I think the red one is bigger because it's higher than all the other ones.

[Students decide to start by filling the red glass. They give directions to the teacher.]

Can someone tell me when I should stop?

[As Shavonne lies flat on the rug, the teacher comments that the girl is positioning herself to get a good look.] **What are you looking at?**

Shavonne: The height.

So she's looking at the height to see if there's a little more space. Is there a little more space?

Shavonne: Yeah, a little bit.

[The teacher adds a bit more water. Another student says that there is a tiny bit more room. The teacher adds a bit more. Another student thinks there is still more room. The teacher adds more and it overflows into the tub.]

Now what?

Jonah: Use the funnel and pour the red into the next cup and see if it overflows.

[The teacher pours the red glass into the green cup.] **It took all this water. This red glass is now empty. Everything that was in here is now all in there. Could I fit any more?**

[Kristi Ann gets up and looks from all angles. She suggests they add "a tiny bit." The teacher adds more water.]

OK, which holds more, the red or the green? First I filled this. Then we poured all this into here, plus a little more. What do you think?

Eva: Green. *[Quite a few hands go up in disagreement.]*

Diego: Red. Because you filled it more times.

Who wants to explain their thinking about which holds more?

Shavonne: I think they might be close because some of the red spilled, and the water you added might have been the same as what you spilled.

Very observant. She thinks maybe we should call them equal or almost equal. Let's move on and try this one [flowered glass]. *[The teacher pours from the green cup into the flowered glass.]*

Eva: It could still use more.

I have an empty green cup here. Eva never had to tell me to stop.

Tuan: It definitely needs more. *[Lots of agreement on this.]*

Luis: Yeah, and it's definite. Not just a little bit.

So which two hold the same?

Iris: The red and the green.

Max: Yeah, because the flowered one isn't filled yet.

Jonah: The flowered one is bigger because it had more space.

So what did we have to do?

Shavonne: We had to add more water. Because it was wider…

Anything to add on?

Diego: The rim is wider. Those cups, if you turn them upside down, the rims of those two are smaller. See? This flowered cup had a wide rim. The space in between [the two rims] is bigger.

INVESTIGATION 3

Measuring Length

What Happens

Session 1: Longer Than, Shorter Than As a class, students compare the lengths of a few objects in the classroom. Then they work individually to compare various objects with a pencil to determine which is longer.

Session 2: Measuring with Hands and Feet Students use their feet and then their hands to measure a strip of tape on the floor. They compare their results and discuss ways of measuring. Then they measure objects in the room using their hands or feet.

Session 3: Feet Lengths Students bring from home cutouts of several different feet they have traced. Working individually, they place their foot outlines in order from smallest to largest (or the reverse). Students then search for foot outlines that are a certain length, measured in lengths of interlocking cubes.

Sessions 4 and 5: Measuring with Cubes Students choose five items to measure with cubes. They list or draw these in order from shortest to longest (or the reverse).

Routines Refer to the section About Classroom Routines (pp. 87–94) for suggestions on integrating into the school day regular practice of mathematical skills in counting, exploring data, and understanding time and changes.

Mathematical Emphasis

- Developing language to describe and compare lengths
- Comparing lengths directly
- Measuring and comparing lengths using nonstandard units
- Ordering lengths
- Representing measurements with numbers, concrete materials, and pictures

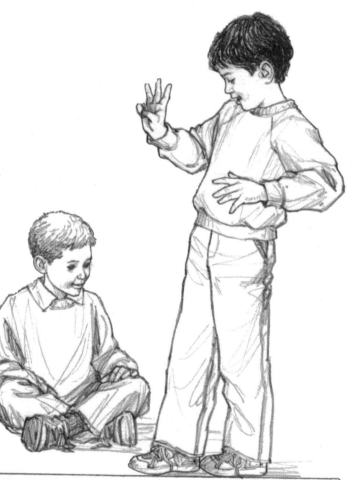

INVESTIGATION 3

What to Plan Ahead of Time

Materials

- Pencils for comparison (these can differ in length): 1 per student (Session 1)
- Masking tape or colored electrical tape (Session 2)
- Interlocking cubes: about 20 per student (Sessions 3–5)
- Half sheets of chart paper (about 16 by 26 inches): 1 per student (Session 3)
- Glue (Session 3)
- Markers or crayons (Sessions 3–5)
- Large sheets of paper (11 by 17 inches): 1 per student (Sessions 4–5)

Other Preparation

- Before Session 1, put together a Measuring collection of 8 items for every 4–5 students. Items should be 2–18 inches long (some shorter, some longer than a pencil). Include items that may be harder to measure (e.g., a coaster), as well as some that are easier to measure (e.g., a small box). Label each item with its name, on masking tape.
- Before Session 2, place a highly visible strip of masking tape or colored electric tape, about 6–8 feet long, on the floor. Also place strips of tape along the full length of about 12–16 classroom items, between 3 and 8 feet long, that students can measure with their hands or feet: bookcases, rug, desk, doorway, table, low windows. Also place a strip of tape along two very short objects, one about a *half* a hand long, the other about a hand and a half.

- Before Session 2, prepare foot outlines of your own feet. Copy and cut out enough to cover the full length of the floor tape, and one or two extra.
- Before Session 3, ask a few students to collect foot outlines from willing people in the school (principal, custodian, art teacher, other teachers). Make multiple copies.
- Duplicate the following student sheets and teacher resources, located at the end of this unit. If you have Student Activity Booklets, copy only those items marked with an asterisk.

For Session 1

Student Sheet 14, Pencil Comparisons (p. 118): 1 per student

Student Sheet 15, Foot Outlines from Home (p. 119): 1 per student

For Session 2

Student Sheet 16, Measuring with Hands and Feet (p. 120): 1 per pair, plus 1 per student for homework in Session 3

Student Sheet 17, Shorter Than My Arm (p. 121): 1 per student

For Session 3

Student Sheet 16, Measuring with Hands and Feet (see Session 2)

Student Sheet 18, Foot Match-Ups (p. 122): 1 per student

Foot Outlines* (p. 124): several, as needed

For Sessions 4 and 5

Student Sheet 19, How Many Cubes Long? (p. 123): 1 per student

100 Charts (p. 126): as needed for reference

Session 1

Longer Than, Shorter Than

Materials

- Measuring collections (8 items per 4–5 students)
- Pencils, for comparison (1 per student)
- Student Sheet 14 (1 per student)
- Student Sheet 15 (1 per student, homework)

What Happens

As a class, students compare the lengths of a few objects in the classroom. Then they work individually to compare various objects with a pencil to determine which is longer. Students' work focuses on:

- understanding what length is
- comparing lengths directly

Activity

Which Is Longer?

Gather students where they can see each other and one of the Measuring collections, perhaps in a circle on the floor or on chairs, with the objects in the center of the circle. Hold up one of the objects.

Suppose I wanted to measure the longest part of this [bottle]. Where would I measure? What's the longest part?

Ask for volunteers to identify the longest part of the object. As students indicate which dimension they think is longest, run your finger along that dimension.

Let's try another one. Suppose I wanted to find out the longest part of this [book]. Where would I measure? Why do you think that is the longest part?

Once again, ask for volunteers to show where to measure. If students have trouble understanding what you mean by "the longest part" of an object, choose a few more especially long items from the collection (or elsewhere in the classroom). For each, ask students to identify the longest dimension.

Now hold up again the first two objects they looked at, one in each hand, holding them as far apart as possible.

I want to know which of these is longer, the [bottle] or the [book]. What do you think? How could you find out for sure?

Some students will suggest that you can tell just by looking at the two things. Encourage them to think about ways they could be sure or could prove it to someone else. Some students might suggest using a ruler or another unit for comparison, such as a hand span. Acknowledge that soon the class will be using other things to measure with, but that first you'd like to find a way to compare the two lengths without using anything else. If students do not suggest directly comparing the lengths by placing the objects together, one long side against the other, raise this idea yourself. Ask how this method could help them determine which is longer.

Now, ask for volunteers to choose any two items from the collection, hold them in separate hands, and ask the rest of the group, "Which is longer? How do you know for sure?" Repeat a few times until students seem to know where to find the longest dimension and have some ways to directly compare two different lengths to see which is longer. Also encourage ideas about using a unit, such as cubes, to measure length. See the **Dialogue Box, Which Is Longer?** (p. 67), for strategies suggested in one classroom.

Activity

Longer and Shorter

Each student needs a pencil (as a tool for comparison) and a copy of Student Sheet 14, Pencil Comparisons. Place each collection of Measuring objects within reach of four or five students and explain that they need to share and take turns with these objects.

Read Student Sheet 14 aloud with the class. As needed, demonstrate on the board how to complete one of the boxes, comparing your pencil and a classroom object.

Here's my pencil, and here's a book. I'll draw the book in the first box. How can I decide which is longer?... The book is longer, so I'll circle the book.

Students work individually. Remind them to compare all eight objects in the collection with their pencil.

Observing the Students

Watch as students make their comparisons.

- Do students figure out which dimension of an object is the longest side? Do they find reasonable ways to measure the length of unusually shaped objects?
- Do students have careful ways of comparing the lengths of two objects? How are they aligning the objects?

As students in each group finish, ask them to compare their results with another student's to see which are the same and which are different. Ask them why they think their results are the same or different.

Note: Students' answers are likely to be different because they have pencils of different lengths. As they compare their results, encourage them to show each other how they measured. If they do not do so on their own, ask them to compare their pencil lengths.

See the **Teacher Note,** Learning about Length (p. 65), for a discussion of issues that may arise during this investigation.

Session 1 Follow-Up

Foot Outlines Distribute Student Sheet 15, Foot Outlines from Home. For this homework, students trace the feet (in socks) of at least four people they know. Read the sheet aloud with them. Point out that they can use any paper available, including old newspaper. Ask for a student volunteer and demonstrate tracing a foot (shoe off) and cutting out the shape. If it is unlikely that students will be able to do this work at home, help them find people in school whose feet they can trace. Students should bring their foot outlines to class by Session 3.

More Pencil Comparisons Students find other objects in the classroom to compare to their pencils. They may record this information on the back of their student sheets.

Learning About Length

Teacher Note

In this unit, students start talking about what's *long, longer, short,* and *shorter*. Their ideas about length begin to develop as they compare lengths directly:

My sister is taller than I am.

My pencil is the shortest in the class.

Research on children's mathematical understanding shows that students typically do not develop a firm idea about length as a stable, measurable dimension until toward the end of the primary grades, although of course there is quite a range of individual differences among students. Through experiences with comparing and ordering lengths, students develop their understanding of what length is and how it can be described.

Students in your class may vary quite a bit in how accurately and consistently they compare the lengths of things. When they begin to measure with objects, such as foot outlines or cubes, you will probably see some who do not carefully line up these objects end to end as they measure, instead either overlapping them or leaving spaces between them. You may also see students who count along the length of an object with the width of a finger or with a single cube, running these along the whole length of the object as they count, but not paying much attention to whether each successive placement begins right where the previous one ended. These "mistakes" are probably not just carelessness or sloppiness; instead, these students are still figuring out what measuring is all about.

Rather than simply telling students to carefully align the ends of two objects in order to compare them, or demonstrating how to measure with a tower of cubes, encourage discussion among students about the different ways they are measuring.

Some people said that this book was 10 cubes long, some said 9, and some said 11. Who would like to show how you measured this book?... OK, William lined up 10 cubes like this. Do you think that's OK? Could it be 11 cubes? Here's a tower of 11 cubes. Could that work? Why or why not?

Continued on next page

Session 1: Longer Than, Shorter Than ■ 65

Teacher Note *continued*

Have fun with this. At times, you might show students some inaccurate ways of measuring in order to help them think through and articulate their own ideas. For example, spread out three cubes, with big gaps, along the edge of a book: one at one edge of the book, one in the middle, and one lined up with the other edge.

Tell students you measured this book and it's three cubes long. Ask them if that seems right and, if it's not, what you should do to get a better measurement. As students discuss and compare ways of measuring, they will gradually develop a sense of what length is and how to measure it accurately.

When students have spent some time comparing lengths directly, they can begin to measure lengths using nonstandard units. Using numbers to measure a continuous quantity is different from using numbers to count a discrete quantity. When students count objects, each successive counting number refers to one object. But to use numbers to measure length, students have to develop the sense of a continuous interval. One unit does not refer to one object, but to an interval. For example, 1 refers to an interval of length from 0 to 1. This is a new and difficult idea. It is important that students develop this idea through many experiences with units that they can place repeatedly and count (such as interlocking cubes, or their feet) so that they physically experience what length is, how it extends from one point to another, and how two lengths might be compared.

Some students may be interested in using rulers. You can certainly make rulers available in the classroom, but in the early grades, students usually see the numbers on the ruler simply as marks to read without understanding much about how a ruler is used as a tool to quantify length. For example, students may align the 1 on the ruler, rather than the end of the ruler, with the edge of an object. They sometimes use the ruler backwards, saying that something is 10 inches when it is actually 2 inches. Students may also think that one kind of measuring unit is equivalent to another; for example, they will find out that something is 6 cubes long and say that the length is "6 inches." They may also think that an object that is longer than another object must be one more unit long. For example, a 7-year-old who knew she was 4 feet tall was comparing herself to a boy in the class. He was a few inches taller, and she said, "So he's 5 feet tall."

As students' understanding of measurement develops over time, they begin to formulate ideas about the need for standard measuring units. Based on the understanding students have developed through many experiences using non-standard units to measure, the use of standard measuring tools becomes meaningful. The need for a standard measuring tool is among the measuring topics students investigate in the grade 3 *Investigations* unit *From Paces to Feet*.

Which Is Longer?

During the activity Which Is Longer? (p. 62), students are trying to figure out which of two objects, a book or a bottle, is longer. Students suggest a variety of comparing strategies, which they describe and try out.

How can we be sure about which is longer, the book or the bottle? People who have tried to tell by looking have been disagreeing.

Eva: Hands. You just put your hands like this *[at the top and bottom of the book]* to measure it. Then you move them to the bottle like this *[keeping them the same distance apart].*

They sure look close. But what happens if you move your hands in midair? It's a good way, but it's easy to make mistakes.

Nadia: A yardstick.

Jonah: Too big.

Claire: You could put them together to see. *[She moves the bottle and the book next to each other, standing them up and lining up the bases of both.]*

Iris: The bottle is taller because the knob is sticking up a little bit.

Jonah: Put your hand across the two tops when they are together. Your hand will be up a little more on the one that is highest.

Yanni: I think they're the same. Where I'm looking, they look the same.

Kristi Ann: Yeah. Because if you put your hand on and if one is more taller, it will tip your hand over.

I think that's like Jonah's idea, but Yanni thought they looked the same. It does look even when I use my hand.

Kristi Ann: How about if you take the cap off?

Well—but let's leave them like this. *[The teacher wants students to decide how the objects compare without changing them.]*

Eva: Use a pencil.

[The teacher rests the pencil across the top, and it stays there. Some students still think they're the same; some think one is taller.]

Nadia: The pencil's going down a little bit on the bottle.

Brady: Close one eye and look like this. *[They do.]*

I'm still having a hard time seeing which one is taller or longer.

Susanna: We could measure it.

What could we use to measure it?

Nathan: Cubes.

Cubes! How tall do you think these would be if we measured with cubes?

[Students' estimates range from 4 to 14. Eva thinks the book will be 11 and the bottle will be 10.]

Tell you what I'm going to do. I'm going to leave these, with the cubes, on the table over there. I want everyone today to take a turn trying to figure out which is longer. You can use the cubes or another method that you think is a good one.

In this example, the bottle and the book are very close in length. Students describe a number of useful techniques for comparing them, including one way that is probably quite familiar. Students often compare heights with each other by lining up back to back and placing a hand or object across the top of the two heads. The teacher encourages students to pursue the comparison question further by giving them the opportunity to handle the bottle and the book directly and to use cubes to compare them.

Session 2

Measuring with Hands and Feet

Materials

- Strip of tape on the floor
- Prepared outlines of teacher's feet
- Student Sheet 16 (1 per pair)
- Student Sheet 17 (1 per student, homework)

What Happens

Students use their feet and then their hands to measure a strip of tape on the floor. They compare their results and discuss ways of measuring. Then they measure objects in the room using their hands or feet. Students' work focuses on:

- using direct comparison of length
- repeating a nonstandard unit to quantify length
- describing ways of measuring length
- describing lengths that can't be measured exactly in whole units

Activity

Measuring a Strip of Tape

Gather students around the strip of tape on the floor. They should sit where they can see it but are not covering it.

A long time ago, when people started measuring the lengths of things, what might they have used?

Students are likely to suggest items such as rocks and sticks; these are certainly possibilities. If no one suggests using part of the body, bring that up yourself.

You could use *yourself* to measure some long things, couldn't you? In fact, long ago, people did use parts of their bodies to measure with, because their bodies—hands, feet, fingers—were always available, no matter where they were or what they wanted to measure. In fact, nowadays people still use their bodies to measure things sometimes.

Today we're going to use our own hands and feet to measure with. We'll start by using our feet. How do you think you could use your feet to measure the length of this strip?

Students may have many ideas about this, including taking off all their shoes and measuring the strip of tape in "shoes." Try a few of their ideas, and use this opportunity to talk about different ways of measuring.

68 ■ *Investigation 3: Measuring Length*

As you introduce the term *length* to describe linear measurement, use this word freely in context so that students hear it often. See, for example, the discussion in the **Dialogue Box,** More Steps or Less? (p. 73).

If the idea of pacing in baby steps, heel to toe, has not yet come up, introduce it.

Some people use "baby steps" to measure length. They go very carefully, putting heel to toe, one foot after the other, like this. *[Demonstrate.]* **If I measure this strip with baby steps, where would I start? Where would I end? Let's have someone try it.**

Ask for a volunteer to use baby steps to pace off the strip.

How many of Leah's feet long is this strip? *[Record this number on the board.]* **What do you think would happen if Leah did it again? Would we get the same number of steps?**

Discuss repeating the measuring process as a way of checking the first measure, and ask the same volunteer to pace the strip again.

The first time Leah did it we got [11] steps, and the second time we got [12] steps. Why do you think that happened? What was different and what was the same about the two times she measured the strip of tape?

Which measurement do you think was more accurate? Why?

If the number wasn't the same both times, talk with students about how they might resolve it. What could have been different about the two different measurements? Were there gaps or overlaps in the volunteer's steps? Did the pacer start and end at the same place each time? During this discussion, keep in mind that measuring is always inexact. There is no "right" way to measure that guarantees an accurate measurement; rather, we look for methods that are as accurate as possible and then try to use those methods carefully.

Make the outlines of your feet available so students can use these to illustrate their ideas about what happened. You may want to demonstrate some reasons for getting different measurements (for example, exaggerate overlapping feet, or begin in the middle of the tape strip).

Students may want to repeat the pacing one more time to make sure. Encourage this, because double-checking measurements is a good habit for students to adopt.

Now let's try it with a different person. Who else wants to measure our strip with baby steps? Do you think our second pacer will get the same number as our first one? Why or why not?

Ask students to explain their reasoning. They may want to compare the feet of the two people directly by lining them up side by side.

So Andre's feet are bigger than Leah's. Do you think he will get a bigger number or a smaller number when he measures this strip with his feet?

Some students may think that a bigger-size foot will lead to a bigger numerical result, while others may be able to describe why a bigger foot results in a smaller number of steps. You might walk the strip yourself to demonstrate an even bigger difference in foot size. Leave a trail of your foot outlines as you demonstrate. Encourage students to talk about this question, but don't expect it to be resolved for all students.

As students are pacing, it is likely that their baby steps will not always end exactly at the end of the tape. When this happens, ask how they would decide what the length is. For example, if the count was 10 but the tenth step extended beyond the end of the tape, what is the length? Some students will probably suggest using 9 steps as the measurement; some may want to use 10 steps; others may want to describe somehow that the measurement is between 9 and 10. Build on ideas that have come up in previous investigations about fractional parts of units. The idea of *half* will likely come up again in the next activity, with more chances to discuss measures of length that fall between whole numbers.

Michelle said that there's an extra half step. So, there are 9 whole steps and a half step more. How could we write this down?

Take suggestions from your students about how to write it. Demonstrate writing it both in words ("9 steps and a half step") and in the correct numerical notation.

Now we're going to try another way of measuring. We're going to measure the length of the strip in *hands*, rather than feet. How could we do this?

Talk about some ways to do this, arriving at a way that is similar to the way you used feet to measure. Ask for volunteers to take "baby steps" with their hands. (Point out where the "heel" of someone's hand is.)

How many of Diego's hands long do you think the strip is? Will it take more "steps" with his hands than if Diego measured with his feet? Which are bigger, his hands or his feet? How could we tell? Let's try it.

Again, ask students to notice details of how the strip was measured. Did the person start at the beginning of the tape? Try the measurement again to see if you get the same results. Did the person do it the same way?

Activity

Measuring Things in the Classroom

In this activity, students will use their feet or their hands to measure the objects in the classroom you have marked with lengths of tape. Before they begin, spend some time on the idea of halves. Using a very short object (less than one hand long) may help students visualize a length that is only part of a unit.

If I wanted to measure the length of this [pack of number cards] with my hand, I'd go like this. *[Place your hand next to the object you have chosen.]* **How many hands long is it?**

Students may say that the object is less than one hand long, a part of a hand, or half a hand.

This is the length of my whole hand *[run your finger along the length of your hand]*, **but where is about half my hand?** *[Ask students to show you.]* **Since this [pack of cards] is about as long as half my hand, we could say it's a *half* a hand long. How could we write this down?**

Again, take students' suggestions and show them the notation for ½.

Then show students an object that is about 1½ hands long (perhaps a large stapler, or a workbook). Ask students how they would describe the length of this object and how you could write its length. Thinking about *half* as a number between two whole numbers is a new idea for many students. Some may say that the object is 2½ hands long; others may say that it is 2 hands, or between 1 and 2 hands.

Since this book is more than 1 hand long by about a half a hand, we can call it 1 and a half hands long.

Write the notation for 1½ on the board. Read this number aloud as you write so that students can see what it looks like. A few students may begin to use the fraction themselves as they continue measuring in this investigation, but don't expect or insist on its use.

Distribute Student Sheet 16 to each pair of students and point out the tape you have placed on objects around the classroom. Explain that they will work in pairs to find the length of each object, using either their hands or their feet.

You can use your feet or hands to measure things on the floor, and your hands to measure things that aren't on the floor. For example, it would be pretty hard to measure the window with your feet, so there you would use your hands.

There are lots of different things to measure, including this strip of tape that Leah and Andre just measured, and also the shorter things that we just measured.

You'll be working in pairs. Take turns doing the measuring while the other person helps and writes. For example, if Nadia measures the bookcase, Chris counts and records. Then switch places when you choose the next thing to measure. Everybody should have a chance to measure some things, and to check and record for other things.

Read through the student sheet with the class, being sure they understand how to record. They write or draw the object measured in the first blank, then write the number that tells how many hands or feet long it is, and finally circle the word that tells which they measured with.

❖ **Tip for the Linguistically Diverse Classroom** As a visual reference, suggest that students sketch a hand and a foot over the first pair of words on their student sheet.

To get the activity started, assign pairs of students to work with each of the taped lengths. When a pair finishes measuring and recording the length of one item, they move on to another.

For the rest of the class period, they measure as many of the objects as they have time for. Student Sheet 16 has a place to record eight lengths. Students may not get to all of those in this class session, but they should measure at least two or three objects with their hands and two or three with their feet.

Session 2 Follow-Up

Shorter Than My Arm For Session 4, students will need to bring in two items from home that they would like to measure. Each item should be non-breakable and shorter than the length of the student's arm. The assignment is described on Student Sheet 17, Shorter Than My Arm.

Note: Remind students to bring in their foot outlines (Student Sheet 15) for the next math class.

More Steps or Less?

These students are about to begin the activity Measuring a Strip of Tape (p. 68). They have agreed to take their shoes off and walk heel to toe, in "baby steps." Jamaar goes first, starting at one end of the tape strip, while the class counts the number of steps aloud together.

So Jamaar found out that the length of the strip was 8 of his own feet. And look where he stopped walking, just about at the end of the tape. Let's try again, with a different person. Will it be the same, also 8?

Leah: No, because people's feet are smaller.

What about my foot?

Fernando: Lower.

My foot would be smaller?

Fernando: No, bigger. It takes more space.

So if I measure this line with my feet, it'll take how many steps?

Fernando: Like 7 or 6. *[The teacher walks the strip, getting 6 steps.]*

Now we're going to measure the length of the strip using our hands. How could we do that?

Michelle: Just like with our feet. *[She demonstrates.]*

Will it take more steps or less?

Leah: More, because your hand is smaller than your foot.

[The teacher asks students to raise their hands if they agree with Leah. A few do, but many think it will take fewer steps, since a hand is smaller than a foot.]

How many of Tony's hands do you think our tape will be?

[Students make a few predictions, then count out loud as Tony measures the tape strip with his hands. When he has finished, some students think the total is 14 and a half, and others call it 15 and a half.]

Let's try again. Remember to start with your hand right at the beginning of the tape.

[Tony repeats; when he takes the extra partial "step," the teacher says "13 and a half."]

Fernando: I thought it was 14.

Claire: I thought it was 15.

[They watch Tony one more time and agree that the number is more than 13 and less than 14. As they compare that to their predictions, they notice that the number of hands is more than the number of feet for the same length.]

Would it make a difference if I measured the length of the strip with my hands? *[Most students say yes.]* **Why?**

Yukiko: Because your hands are different.

How are they different?

Nadia: They're bigger.

How much bigger?

Fernando: A little bit. So it'd be like 10.

The teacher measures and finds the length is a little bit more than 10 hands. Throughout this discussion, students have been involved in thinking about how to repeat a unit to measure a length, and how different units give different counts for the same length.

Session 3

Feet Lengths

Materials

- Foot outlines brought from home
- Extra foot outlines (optional)
- Interlocking cubes (at least 20 per student)
- Chart paper, half sheets (1 per student)
- Glue (to share)
- Markers (to share)
- Student Sheet 18 (1 per student)
- Student Sheet 16 (1 per student, homework)

What Happens

Students bring from home cutouts of several different feet they have traced. Working individually, they place their foot outlines in order from smallest to largest (or the reverse). Students then search for foot outlines that are a certain length, measured in lengths of interlocking cubes. Their work focuses on:

- putting several different lengths in order (seriating)
- matching lengths
- comparing lengths

Activity

Putting Feet in Order

At home, you've been tracing and cutting out feet of different lengths. Let's see what you have.

As students show their foot outlines, be sure that each is labeled with the contributing person's name. Each student should have four different outlines. If any do not have enough, they can supplement their collection with foot outlines collected at school or with copies of the foot outlines provided with this unit (p. 124). Ask for volunteers to hold up their foot outlines one by one and tell the owner of each foot.

Today you're going to make a foot poster. You'll be putting your foot outlines in order *by length*. What do you think that means? How would you do it?

Students share ways to order their outlines. For example, they might suggest starting with their own foot, then finding the feet that are longer and shorter. Or they might start by finding the longest foot and working down to the shortest, or the reverse.

Suppose we want to go shortest to longest. Where would we start? How would we figure out which foot is the shortest?

74 ■ *Investigation 3: Measuring Length*

Take four foot outlines from your own collection, or one each from four different students, and a half sheet of chart paper. Ask for a volunteer to describe how to start. Encourage discussion of how to place the feet for comparing their lengths, and how to make sure you've identified the shortest one. There may also be some discussion about where to measure the length of the foot, because some people have longer big toes and some people have longer second toes.

What if we discover two feet that are the same length?

Ask students for suggestions about how to place these on a poster or how to label them, so that it's clear that they're the same length.

After you have your four foot outlines in order of size, you're going to glue them on your big piece of paper to make your poster. As soon as you're sure you have them in order, from short to long or long to short, check in with me. Then you're ready to glue them down. Try to place them on your paper so it's clear to someone looking at your poster how the feet go from longest to shortest, or from shortest to longest.

Distribute the large sheets of paper. Have markers available for students who want to label their posters. If some students finish quickly, you might provide additional foot outlines for them to add to their collection. Provide the glue to students as they check in with you.

Observing the Students

As students check in with you, ask them to explain how they ordered their foot outlines (from shortest to longest, or longest to shortest) and how they know that the feet are all in the proper order. Watch for the following:

- How do students align the feet to compare the lengths? Do their strategies lead to accurate results?
- Do students compare carefully, especially when two measurements are close to each other?
- Do students understand how to compare the feet systematically? For example, do they repeat comparisons that have already been made? Do they use information from a previous comparison in order to help with a new comparison?
- Can students explain their strategy for ordering? How do they know for sure that they have the feet in order?

If students are having difficulty ordering their feet, suggest that they start by finding the shortest one and the longest one.

When students have finished, they post their papers where other students can see and reach them for the next activity.

Activity

Foot Match-Ups

As students finish their posters, give each one a copy of Student Sheet 18, Foot Match-Ups, and about 20 interlocking cubes. Briefly explain the directions.

When you've finished putting your foot outlines in order, you're going to start looking for certain feet on our posters. There are six feet listed on this sheet by their length. It tells how many cubes long they are.

Read aloud the six lengths listed on the sheet: 6 cubes long, 9 cubes long, 12 cubes long, 14 cubes long, less than 6 cubes long, more than 14 cubes long.

Suppose, for example, you start out looking for the foot that is 6 cubes long. First you need to see how long "6 cubes" is.

Demonstrate making a tower 6 cubes long and looking among the posters for a foot outline that matches the tower in length. Explain that you are looking for a foot that is close to 6 cubes long, because the match-up doesn't have to be absolutely perfect.

When you've found a match for a certain length on your sheet, write the name of the person you see on that foot. There may be more than one good match, so if you'd like, write as many as you can find. For some lengths, there may not be a match at all.

For the rest of the class period, students work on finding match-ups between the foot outlines and the lengths listed on Student Sheet 18. If they seem to need help with what "less than 6 cubes long" and "more than 14 cubes long" mean, you might discuss this with the whole class. As more students finish their foot posters, there will be more opportunities for everyone to find foot match-ups.

Some students may not get a chance to try the foot match-ups during this session, and many will not be able to finish. You may want to give them time during another session or at other times during the day to finish their matching work.

Session 3 Follow-Up

Measuring with Hands and Feet Send home an additional copy of Student Sheet 16, Measuring with Hands and Feet. Students find objects at home to measure in the same way they did in class, using either hands or feet.

Length Book Read *Math Counts: Length* by Henry Pluckrose (Childrens Press, 1995) with your class. Point out the ways of measuring used in this book that are similar to the ones you used in class (hand spans and paces).

More About Halves Read *The Half-Birthday Party* by Charlotte Pomerantz (Clarion Books, 1984). In this story, a little boy organizes a "half" birthday party for his 6-month-old baby sister. Help students understand each "half" gift that someone brings to the party. Students then brainstorm their own "half" gifts to create a class book of halves.

Sessions 4 and 5

Measuring with Cubes

Materials

- Large paper, 11 by 17 inches (1 per student)
- Interlocking cubes (available in sets of 30–40 per pair)
- Markers or crayons
- Measuring collections (from Session 1)
- Objects from home (2 per student)
- Student Sheet 19 (1 per student)
- Unlined paper (1 sheet per student)
- 100 charts (available for use as needed)

What Happens

Students choose five items to measure with cubes. They list or draw these in order from shortest to longest (or the reverse). Their work focuses on:

- comparing and seriating (ordering) lengths
- repeating a nonstandard unit along a length
- representing measurements in a clear, ordered way
- presenting information clearly to others

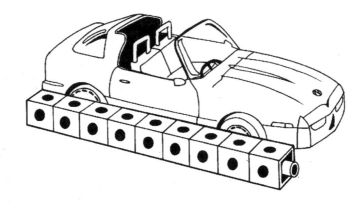

Activity

Measuring Objects with Cubes

Begin this session by asking students to take out the two objects they brought from home to measure in class.

Today you will work with a partner to measure the things you brought in from home, as well as some other objects from our Measuring collections. This time we are going to use *cubes* to measure with, the way you did for the Foot Match-Ups.

Distribute Student Sheet 19, How Many Cubes Long? to each student. Every group should have a supply of about 30–40 interlocking cubes available.

Pick one object to start with. How long is it? Measure the longest part of it in cubes. *[Demonstrate as needed.]* When you and your partner agree on how many cubes long your first object is, record the number on this sheet. Use pictures and words to describe your object.

If you're measuring your panda bear, you write "panda" or "bear" on the first line. Then you draw a little panda. Finally, you write how many cubes long the panda is. If the object you're measuring is from the class collection, be sure to put it back so someone else can use it.

Bring up the issue of partial units.

What if you find that an object is in between 4 cubes and 5 cubes long. How could you decide how long it is?

Students may decide to choose whichever number is closer. They may suggest remeasuring. Or they may bring up the idea of halves or "extras." All are reasonable options. Ask students for different ways to write a length between 4 and 5. Include the standard notation for 4½ to remind students of this possibility for a length that seems to be midway between the two.

Observing the Students

Watch while students measure and record the lengths of five objects.

- Do students select the longest dimension to measure?
- Are students aligning the beginning and end of a cube tower with the object they are measuring, or do they forget to align one or both of these?
- Are students checking again when they aren't sure of their measurements?
- Are students counting and keeping track of the number of cubes correctly?
- What do students decide to do when a length is not close to a number of whole cubes?

If some students finish earlier than others, they can continue measuring other objects in the classroom shorter than an arm's length or finish their work on the Foot Match-Ups.

Activity

When most students have finished recording their five measurements, call the class together to explain the follow-up assessment activity. Each student will need a sheet of unlined paper. Have 100 charts available for students to refer to if needed.

Each of you will get a sheet of paper. On this sheet, you need to find a way to show your objects in order from shortest to longest, or from longest to shortest.

Assessment

Representing Objects in Order

First of all, how can you decide which object is the shortest?

Students may suggest looking at the numbers of cubes they've written on Student Sheet 19, to see which is the smallest number. Or, they may build and directly compare cube towers of the recorded lengths. Some students may want to retrieve the actual objects to compare. If they suggest relying on memory or just looking, ask them to also do it another way to be sure.

What are some ways you could show your objects in order on a sheet of paper?

Students may suggest drawing their objects on the page in some organized way (from top to bottom, or left to right). If it is not obvious from the size of their drawings, students might want to write the length of the objects in cubes (for example, 12, 9, 8, 6, 3) next to the drawings. Instead of drawing, students might list their objects by name, in order, including each measurement.

You may use pictures, numbers, and words, or any combination of those. Just be sure that it's clear to someone else how your objects are ordered. When you think you have finished, check with someone else to see if they can tell from your paper which of your objects is the shortest and which is the longest.

Students will probably spend the rest of Session 4 and part of Session 5 making their representations. If some students need help deciding which numbers are larger or smaller than others, they can use a 100 chart as a reference.

Observing the Students

As students are working, watch for the following:

- What strategies are students using to decide how to order their objects? Do they use numbers to help them? Do they compare cube towers? the objects themselves? Do they double-check their work?

- How are students' representations organized? Do they clearly show their five objects in order by length? Do they use the layout of the page, the size of drawn objects, numbers, or words to show their order?

See the **Teacher Note,** Assessment: Representing Objects in Order (p. 82), for further discussion of students' work on this activity.

Activity

Sharing Representations

At the end of Session 5, ask for two or three students to share their representations with the class.

Looking at these representations, can you tell which is the shortest object in each picture? which is the longest? Can you tell which is the *second* longest object?

What's the same about these representations? What's different? What helps you understand them?

Ask students to describe aspects of the representations that help them understand the ordering of the objects. Then call on another two or three students and repeat this process until the end of the session, or until everyone who wants to share has done so.

Display the representations in the classroom so students can look at each other's work more carefully.

Activity

Choosing Student Work to Save

As the unit ends, you may want to use one of the following options for creating a record of students' work on this unit.

- Students look back through their work and think about what they learned in this unit, what they remember most, what was hard or easy for them. You might have students discuss this with a partner or have students share in the whole group.

- Depending on how you organize and collect student work you may want to have students select some examples of their work to keep in a math portfolio. In addition you may want to choose some examples from each student's folder to include. Items such as their work on the final assessment, Representing Objects in Order (p. 79), can be useful pieces for assessing student growth over the school year.

- You may want to send a selection of work home for parents to see. Some teachers have found it helpful to include a short letter to parents summarizing the work in this unit. You could enlist the help of your students and together generate a letter that describes the mathematics that students were involved in. If you are keeping a year-long portfolio of mathematics work for each student this work should be returned to you.

Teacher Note: Assessment: Representing Objects in Order

For the final assessment, students make representations of five objects ordered by length. One teacher sorted her students' work into three groups to help her consider how they were understanding linear measure. Besides what she could see in the representations, she thought about what she had seen while the students worked on this task.

One group of students made clear representations of five objects ordered by length. For example, Kaneisha and Luis used words and numbers to list their objects in order, while Eva used pictures and numbers. Many other students approached the task in similar ways, using some combination of words of pictures with numbers.

A second group of students were not quite so fluent, but were able to work through the difficulties they had and create a complete solution. Garrett drew five objects and wrote their lengths, but neglected to arrange the objects in order on the page as he went along. However, once he had all five objects on the page, he added arrows to indicate their order.

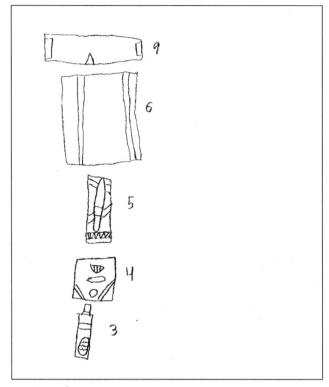

Eva's work

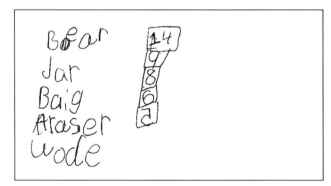

Kaneisha's work

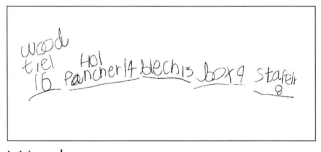

Luis's work

Garrett's work

82 ■ *Investigation 3: Measuring Length*

Teacher Note
continued

Susanna's work

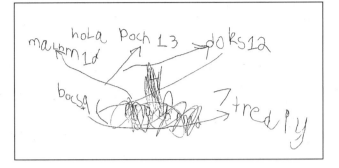

Libby's work

Susanna's completed paper shows only four objects in order. However, she measured very carefully and thought hard about how to record numbers in between whole cubes. As the teacher watched her measure the eraser, Susanna said, "There's a little bit sticking off here. It's about a half, so I'll say 7 and a half."

The teacher felt that Garrett, Susanna, and others who did similar work were, like those in the first group, well on their way to using units to measure and compare length.

A few students fell into a third group; they were still struggling with some of the beginning ideas about comparing length. Typically, they could measure objects with the cubes, but could not order four or five objects.

Libby competently measured eight objects and recorded their names and lengths. However, she had a great deal of trouble putting five in order. She would choose an object that she thought was the longest and write down its name and length. Then she might find another object that was shorter than the first and would record it somewhere on her paper. When she picked up an object that fell between those two, she would not know where to put it.

Like Libby, Chris was not able to order five objects. While he measured and recorded the lengths of quite a few objects, he was not able to read what he had recorded and could not even order five objects by using the objects themselves. The teacher asked him to find a way to say something about the order of a few of his objects. He wrote the following:

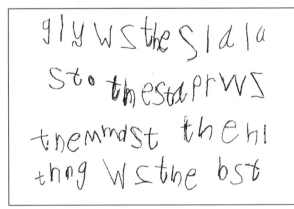

"Glue was the smallest. The stapler was the mediumest. The hole thing [puncher] was the biggest."

The teacher realized that these students were comfortable directly comparing two or three objects, but were not sure how the numbers of cubes could help them put objects in order. She decided that these students would need more opportunities during the school year to continue measuring length and comparing the lengths of two or three objects.

Teacher Note: About Choice Time

Choice Time is an opportunity for students to work on a variety of activities that focus on similar mathematical content. Choice Times are found in every unit of the grade 1 *Investigations* curriculum. These generally alternate with whole-class activities in which students work individually or in pairs on one or two problems. Each format offers different learning experiences; both are important for students.

In Choice Time the activities are not sequential; as students move among them, they continually revisit some of the important concepts and ideas they are learning. Many Choice Time activities are designed with the intent that students will work on them more than once. As they play a game a second or third time, or as they work to solve similar problems, students are able to refine their strategies, see a variety of approaches, and bring new knowledge to familiar experiences.

You may want to limit the number of students working on a particular Choice Time activity at any one time. In many cases, the quantity of materials available limits the number. Even if this is not the case, limiting the number is advisable because it gives students the opportunity to work in smaller groups. It also gives them a chance to do some choices more than one time. Often when a new choice is introduced, many students want to do it first. Assure them that, even with your limits, they will have the chance to try each choice.

Initially you may need to help students plan what they do. Rather than organizing them into groups and circulating the groups every 15 minutes, support students in making their own decisions. Making choices, planning their time, and taking responsibility for their own learning are important aspects of a student's school experience. If some students return to the same activity over and over again without trying other choices, suggest that they make a different first choice and then do the favorite activity as a second choice.

How to Set Up Choices

Some teachers prefer to have the choices set up at centers or stations around the room. At each center students will find the materials needed to complete the activity. Other teachers prefer to have materials stored in a central location, with students taking the materials to their own desks or tables. In either case, materials should be readily accessible, and students should be expected to take responsibility for cleaning up and returning materials to their appropriate storage locations. Giving a "5 minutes until cleanup" warning before the end of any session allows students to finish what they are working on and prepare for the upcoming transition.

Decide which arrangement to use in your classroom. You may need to experiment with a few different structures before finding the setup that works best for you and your students.

The Role of the Student

Establish clear guidelines when you introduce Choice Time. Discuss students' responsibilities:

- Try every choice at least once.
- Work with a partner or alone. (Some activities require that students work in pairs, while others can be done either alone or with a partner.)
- Keep track, on paper, of the choices you have worked on.
- Keep all your work in your math folder.
- Ask questions of other students when you don't understand or feel stuck. (Some teachers establish the rule, "Ask two other students before me," requiring students to check with two peers before coming to the teacher for help.)

For each Choice Time, list the activity choices on a chart, the board, or the overhead. Sketch a picture with each choice for students who may have difficulty reading the activity names. Some teachers laminate a piece of tagboard to create a

Choices board that they can easily update as new choices are added from session to session and old choices are no longer offered.

First grade students can keep track of the choices they have completed in one of these ways:

- When they have completed an activity, students record its name or picture on a blank sheet of paper.
- Post a sheet of lined paper at each station, or a sheet for each choice at the front of the room. At the top of each sheet, put the name of one activity and the corresponding picture. When students have completed an activity, they print their name on the corresponding sheet. Keep these lists throughout an investigation, as the same choices may be offered several times.

Some teachers keep a date stamp at each Choice Time station or at the front of the room, making it easy for students to record the date as well.

In any classroom there will be a range of how much work students complete. Some choices include extensions and additional problems for students who have completed their required work. Encourage students to return to choices they have done before, do another problem or two from the choice, or play a game again. You may also want to make the choices available at other times during the day.

Whenever students do any work on paper during Choice Time, they put this in their math folders at the end of the session.

At the end of a Choice Time session, spend a few minutes discussing with students what went smoothly, what sorts of issues arose and how they were resolved, and what students enjoyed or found difficult about Choice Time. Having students share the work they have been doing often sparks interest in an activity. Some days, you might ask two or three volunteers to talk about their work. On other days, you might pose a question that someone asked you during Choice Time, so that other students might respond to it. Encourage students to be involved in the process of finding solutions to problems that come up in the classroom. In doing so, they take some responsibility for their own behavior and become involved with establishing classroom policies.

The Role of the Teacher

Choice Time provides you with the opportunity to observe and listen to students while they work. At times, you may want to meet with individual students, pairs, or small groups who need help. This gives you the chance to focus on students you haven't had a chance to observe before, or to do individual assessments. Recording your observations of students will help keep you aware of how they are interacting with materials and solving problems. The **Teacher Note,** Keeping Track of Students' Work (p. 86), offers some strategies for recording and using your observations.

During the initial weeks of Choice Time, much of your time will probably be spent in classroom management, circulating around the room, helping students get settled into activities, and monitoring the process of moving from one choice to another. Once routines are familiar and well established, students will become more independent and responsible for their work during Choice Time. This will allow you to spend more concentrated periods of time observing the class as a whole or working with individuals and small groups.

Teacher Note

Keeping Track of Students' Work

Throughout the *Investigations* curriculum, there are numerous opportunities to observe students as they work. Teacher observations are an important part of ongoing assessment. A single observation is like a snapshot of a student's experience with a particular activity, but when considered over time, a collection of these snapshots provides an informative and detailed picture of a student. Such observations can be useful in documenting and assessing student's growth, as well as in planning curriculum. They offer important sources of information when preparing for parent conferences or writing student reports.

The way you observe students will vary throughout the year. At times you may be interested in particular problem-solving strategies that students are developing. Other times, you might want to observe how students use or do not use materials for solving problems. You may want to focus on how students interact when working in pairs or groups. Or you may be interested in noting the strategy that a student uses when playing a game during Choice Time. Class discussions also provide many opportunities to take note of student ideas and thinking.

You will probably need some sort of system to record and keep track of your observations. While a few ideas and suggestions are offered here, it's important to find a recordkeeping system that works for you. All too often, keeping observation notes on a class of 28–32 students can quickly become overwhelming and time-consuming.

A class list of names is one convenient way of jotting down your observations. Since the space is somewhat limited, it is not possible to write lengthy notes; however, over time, these short observations provide important information.

Another common approach is to keep a supply of adhesive address labels on clipboards around the room. After taking notes on individual students, you can peel off each label and stick it in the appropriate student's file.

Some teachers keep a loose-leaf notebook with a page for each student. When something about a student's thinking strikes them as important, they jot down brief notes and the date.

You may find that writing notes at the end of each week works well for you. Some teachers find this a useful way of reflecting on individual students, on the curriculum, and on the class as a whole. Planning for the next week's activities often grows out of these weekly reflections.

In addition to your own notes, you will have each student's folder of work for the unit. This documentation of their experiences can help you keep track of your students, assess their growth over time, and communicate this information to others. An activity at the end of each unit, Choosing Student Work to Save, suggests particular pieces of work you might keep in a portfolio of work for the year.

Counting

Counting is an important focus in the grade 1 *Investigations* curriculum, as it provides the basis for much of mathematical understanding. As students count, they are learning how our number system is constructed, and they are building the knowledge they need to begin to solve numerical problems. They are also developing critical understandings about how numbers are related to each other and how the counting sequence is related to the quantities they are counting.

Counting routines can be used to support and extend the counting work that students do in the *Investigations* curriculum. As students work with counting routines, they gain regular practice with counting in familiar classroom contexts, as they use counting to describe the quantities in their environment and to solve problems based on situations that arise throughout the school day.

How Many Are Here Today?

Since you must take attendance every day, this is a good time to look at the number of students in the classroom in a variety of ways.

Ask students to look around and make an estimate of how many are here today. Then ask them to count.

At the beginning of the year, students will probably find the number at school today by counting each student present. To help them think about ways to count accurately, you can ask questions like these:

How do we know we counted accurately? What are different methods we could use to keep track and make sure we have an exact count? (For example, you could count around a circle of seated students, with each student in turn saying the next number. Or, all students could start by standing up, then sit down in turn as each says the next number.)

Is there another way we could count to double-check? (For example, if you counted around the circle one way, you could count around the circle the other way. If you are using the standing up/sitting down method, you could recount in a different order.)

You might want to count at other times of the day, too, especially when several students are out of the room. For example, suppose groups of students are called to the nurse's office for hearing examinations. Each time a new group of students leaves, you might ask the class to look around and think about how many students are in the room now:

So, this time Diego's table and Mia's table both went to the nurse. Usually we have 28 students here. Look around. What do you think? Don't count. Just tell me about how many students might be here now. Do you think there are more than 5? more than 10? more than 20?

Later in the year, some students may be able to use some of the information they know about the total number of students in the class and how many students are absent to reason about the number present. For example, suppose 26 students are in class on Monday, with 2 students absent. On Tuesday, one of those students comes back to school. How many students are in class today? Some students may still not be sure without counting from one, but other students may be able to reason by counting on or counting back, comparing yesterday and today. For example, a student might solve the problem in this way:

> "Yesterday we had 26 students, and Michelle and Chris were both absent. Today, Chris came back, so we have one more person, so there must be 27 today."

Another might solve it this way:

> "Well we have 28 students in our class when everyone's here. Now only Michelle is absent, so it's one less. So it's 27."

About Classroom Routines

From time to time, you might keep a chart of attendance over a week or so, as shown below. This helps students become familiar with different combinations of numbers that make the same total. If you have been doing any graphing, you might want to present the information in graph form.

Day	Date	Present	Absent	Total
Monday	March 2	26	2	28
Tuesday	March 3	27	1	28
Wednesday	March 4	27	1	28
Thursday	March 5	27	1	28
Friday	March 6	28	0	28
Monday	March 9	28	0	28
Tuesday	March 10	26	2	28
Wednesday	March 11	25	3	28

After a week or two, look back over the data you have collected. Ask questions about how things have changed over time.

In two weeks of attendance data, what changes? What stays the same?

On which day were the most students here? How can you tell? Which day shows the least students here? What part of the [chart] gives you that information?

Another idea (for work with smaller numbers) is to keep track of the number of girls and boys present and absent each day. Again, many students will count by 1's. Later in the year, some will also reason about these numbers:

> "There are two people absent today and they're both girls. We usually have 14 girls, and Kaneisha's sick, that's 13, and Claire's sick, that's 12."

Can Everyone Have a Partner?

Attendance can be an occasion for students to think about making groups of two:

We have 26 students here today. Do you think that everyone can have a partner if we have 26 students?

Students can come up with different strategies for solving this problem. They might draw 26 stick figures, then circle them in 2's. They could count out 26 cubes, then put them together in pairs. They might arrange themselves in 2's, or count by 2's.

At the beginning of the year, many of your students will need to count by 1's from the beginning each time you add two more students, but gradually some will begin to notice which numbers can be broken up into pairs:

> "I know 13 doesn't work, because you can do it with 12, and 13's one more, so you can't do it."

Some students will begin to count by 2's, at least for the beginning of the counting sequence. Then, as the numbers get higher, they may still be able to keep track of the 2's, but need to count by 1's:

> "So, that's 2, 4, 6, 8, 10, 12, um, 13, 14 . . . 15, 16."

As you explore 2's with your students, keep in mind that many of them will need to return to 1's as a way to be sure. Even though some students learn the counting sequence 2, 4, 6, 8, 10, 12 . . . by rote, they may not connect this counting sequence to the quantities it represents at each step.

One teacher found a way to help students develop meaning for counting by 2's. She took photographs of each student, backed them with cardboard, then used them during the morning meeting as a model for making pairs. She laid out the photos in two columns, and asked about the new total after the addition of each pair:

We have 10 photos out so far. The next two photos are for William and Yanni. When we put

those two photos down, how many photos will we have?

Lining up is another time to explore making pairs. Before lining up, count how many students are in class (especially if it's different from when you took attendance). Ask students whether they think the class will be able to line up in even pairs. For many first grade students, the whole class is too many people to think about. You can ask about smaller groups:

What if Kristi Ann's table lines up first? Do you think we could make even partners with the people at that table?

What about Shavonne's table? ... Do you think Shavonne's table will have an extra? How do you know?

Is there another table that would have an extra that we could match up with the extra person from Shavonne's table?

Once students are lined up in pairs, they can count off by 2's. Because most first graders will need to hear all the numbers to keep track of how the counting matches the number of people, ask them to say the first number in the pair softly and the second one loudly. Thus the first pair in line can say, "1, **2**," the second pair can say, "3, **4**," and so forth.

Counting to Solve Problems

Be alert to classroom activities that lend themselves to a regular focus on solving problems through counting. Use these situations as contexts for counting and keeping track, estimating small quantities, breaking quantities into parts, and solving problems by counting up or back. For example, take a daily milk count:

Everyone who is buying milk today stand up. Without counting yet, who has an idea how many students might be standing up? Is it more than 5? more than 10? more than 50? ... Now, let's count. How could we keep track today so that we get an accurate count?

You can make a problem out of lunch count:

We found out that 23 students are buying school lunch today. We have 27 students here. So how many students brought their own lunch from home today?

Watch for the occasional sharing situation:

Claire brought in some cookies she made to share for snack. She brought 36 cookies. Is that enough for everyone to have one cookie, including me and our student teacher? Oh, and Claire wants to invite her little brother to snack. Do we have enough for him, too? Will there be any cookies left over?

The sharing of curriculum materials can also be the basis of a problem:

Each pair of students needs a deck of number cards to share. While I'm getting things together, work on this problem with your partner. We said this morning that we have 26 students here. If I need one deck for each pair, how many decks do I need?

Exploring Data

Through data routines at grade 1, students gain experience working with categorical data—information that falls into categories based on a common feature (for example, a color, a shape, or a shared function). The data routines specifically extend work students do in the *Investigations* curriculum. The Guess My Rule game and its many variations (introduced in the unit *Survey Questions and Secret Rules*) can be used throughout the year for practice with organizing sets into categories and finding ways to describe those categories—a fundamental part of analyzing data. Students can also practice collecting and organizing categorical data with quick class surveys that focus on their everyday experiences; this practice supports the survey-taking they do in the curriculum.

Guess My Rule

Guess My Rule is a classification game in which players try to figure out the common characteristic, or attribute, of a set of objects. To play the game, the rule maker (who may be the teacher, a student, or a small group) decides on a secret rule for classifying a particular group of things. For example, a rule for classifying people might be WEARING STRIPES.

The rule maker (always the teacher when the game is first being introduced) starts the game by giving some examples of objects or people who fit the rule. The guessers then try to find other items that fit the same rule. Each item (or person) guessed is added to one of two groups—either *does fit* or *does not fit* the rule. Both groups must remain clearly visible to the guessers so they can make use of all the evidence as they try to figure out the rule.

Emphasize to the players that "wrong" guesses are as important as "right" guesses because they provide useful clues for finding the rule. When you think most students know the rule, ask for volunteers to share their ideas with the class.

Once your class is comfortable with the activity, students can choose the rules. Initially, you may need to help students choose appropriate rules.

Guess my Rule with People When sorting people according to a secret rule, always base the rule on just one feature that is clearly visible, such as WEARING A SHIRT WITH BUTTONS, or WEARING BLUE. When students are choosing the rule, they may choose rules that are too obvious (such as BOY/GIRL), so vague as to apply to nearly everyone (WEARING DIFFERENT COLORS), or too obscure (HAS AN UNTIED SHOELACE). Guide and support students in choosing rules that work.

Guess My Rule with Objects Class sets of attribute blocks (blocks with particular variations in size, shape, color, and thickness) are a natural choice for Guess My Rule. You can also use collections of objects, such as sets of keys, household container lids, or buttons. One student sorts four to eight objects according to a secret rule. Others take turns choosing an object from the collection that they think fits the rule and placing it in the appropriate group. If the object does not fit, the rule maker moves it to the NOT group. After several objects have been correctly placed, students can begin guessing the rule.

Guess My Object Once students are familiar with Guess My Rule, they can use the categories they have been identifying to play another guessing game that also involves thinking about attributes. In this routine, students guess, by the process of elimination, which particular one of a set of objects has been secretly chosen. This works well with attribute blocks or object collections.

To start, place about 20 objects where everyone can see them. The chooser secretly selects one of the objects on display, but does not tell which one (you may want the chooser to tell you, privately). Other students ask yes-or-no questions, based on attributes, to get clues to help them identify the chosen object. After each answer, students move to one side the objects that have been eliminated. That is, if someone asks "Is it round?" and the answer is yes, all objects that are *not* round are moved aside.

Pause periodically to discuss which questions help eliminate the most objects. For example, "Is it this one?" eliminates only one object, whereas "Is it red?" may eliminate several objects. For more challenge, students can play with the goal of identifying the secret object with the fewest questions.

Quick Surveys

Class surveys can be particularly engaging when they connect to activities that arise as a regular part of the school day, and they can be used to help with class decisions. As students take surveys and analyze the results, they get good practice with collecting, representing, and interpreting categorical data.

Early in first grade, to keep the surveys quick and the routine short, use questions that have exactly two possible responses. For example:

Would you rather go outside or stay inside for recess today?

Will you drink milk with your lunch today?

Do you need left-handed or right-handed scissors?

As the school year progresses, you might include some survey questions that are likely to have more than two responses:

Which of these three books do you want me to read for story time?

Who was your teacher last year?

Which is your favorite vegetable growing in our class garden?

How old are you?

In which season were you born?

Try to choose questions with a predictable list of just a few responses. A question like "What is your favorite ice cream flavor?" may bring up such a wide range of responses that the resulting data is hard to organize and analyze.

As students become more familiar with classroom surveys, invite the class to brainstorm questions with you. You may decide to avoid survey questions about sensitive issues such as families, the body, or abilities, or you might decide to use surveys as a way of carefully raising some of these issues. In either case, it is best to avoid questions about material possessions ("Does your family have a car?").

Once the question is chosen, decide how to collect and represent data. Be sure to vary the approach. One time, you might collect data by recording students' responses on a class list. Another time, you might take a red interlocking cube for each student who makes one response, a blue cube for each student who makes the other response. Another time, you might draw pictures. If you have prepared Kid Pins and survey boards for use in *Mathematical Thinking at Grade 1,* these can be used for collecting the data from quick surveys all year.

Initially, you may need to help students organize the collected data, perhaps by stacking cubes into "bars" for a "graph," or by making a tally. Over time, students can take on more responsibility for collecting and organizing the data.

Always spend a little time asking students to describe, compare, and interpret the data.

What do you notice about these data?

Which group has the most? the least? How many more students want [recess indoors today]?

Why do you suppose more would rather [stay inside]? Do you think we'd get similar data if we collected on a different day? What if we did the same survey in another class?

About Classroom Routines

Understanding Time and Changes

These routines help students develop an understanding of time-related ideas such as sequencing of events, understanding relationships among time periods, and identifying important times in their day.

Young students' understanding of time is often limited to their own direct experiences with how important events in time are related to each other. For example, explaining that an event will occur *after* a child's birthday or *before* an important holiday will help place that event in time for a child. Similarly, on a daily basis, it helps to relate an event to a benchmark time, such as *before* or *after* lunch. Both calendars and daily schedules are useful tools in sequencing events over time and preparing students for upcoming events. These routines help young students gain a sense of basic units of time and the passage of time.

Calendar

The calendar is a real-world tool that people use to keep track of time. As students work with the calendar, they become more familiar with the sequence of days, weeks, and months, and the relationships among these periods of time. Calendar activities can also help students become more familiar with relationships among the numbers 1–31.

Exploring the Monthly Calendar At the start of each month, post the monthly calendar and ask students what they notice about it. Some students might focus on arrangement of numbers or total number of days, while others might note special events marked on the calendar, or pictures or designs on the calendar. All these kinds of observations help students become familiar with time and ways that we keep track of time. You might record students' observations and post them near the calendar.

As the year progresses, encourage students to make comparisons between the months. Post the calendar for the new month next to the calendar for the month just ending and ask students to share their ideas about how the two calendars are similar and different.

Months and Years To help students see that months are part of a larger whole, display the entire calendar year on a large sheet of paper. Cut a small calendar into individual monthly pages and post the sequence of months on the wall. You might decide to post the months according to the school year, September through August, or the calendar year January through December. At the start of each month, ask students to find the position of the new month on the larger display. From time to time, you might also use this display to point out dates and distances between them as you discuss future events or as you discuss time periods that span a month or more. (Last week was February vacation. How many weeks until the next vacation?)

How Many More Days? Ask students to figure out how long until special events, such as birthdays, vacations, class trips, holidays, or future dates later in the month. For example:

Today is October 5. How many more days until October 15?

How many more days until [Nathan's] birthday?

How many more days until the end of the month?

Ask students to share their strategies for finding the number of days. Initially, many students will count each subsequent day. Later, some students may begin to find their answers by using their growing knowledge of calendar structure and number relationships:

> "I knew there were three more days in this row and I added them to the three days in the next row. That's 6 more days."

Others may begin using familiar numbers such as 5 or 10 in their counting:

> "Today is the 5th. Five more days is 10, and five more is 15. That's 10 more days until October 15."

For more challenge, ask for predictions that span two calendar months. For example, you might post the calendar for next month along side of the calendar of this month and ask a question like this:

It's April 29 today. How many more days until our class trip on May 6?

Note that we can refer to a date either as October 15 or as the 15th day of October. Vary the way you refer to dates so that students become comfortable with both forms. Saying "the 15th day of October" reinforces the idea that the calendar is a way to keep track of days in a month.

How Many Days Have Passed? Ask questions that focus on events that have already occurred:

How many days have passed since [a special event]? since the weekend? since vacation?

Mixed-Up Dates If your monthly display calendar has date cards that can be removed or rearranged, choose two or three dates and change their position on the calendar so that the numbers are out of order. Ask students to fix the calendar by pointing out which dates are out of order.

Groups of two or three can play this game with each other during free time. Students can also remove all the date cards, mix them up, and reassemble the calendar in the correct order. You might mark the space for the first day of the month so that students know where to begin.

Daily Schedule

The daily schedule narrows the focus of time to hours and shows students the order of familiar events over time. Working with schedules can be challenging for many first graders, but regular opportunities to think and talk about the idea will help them begin predicting what comes next in the schedule. They will also start to see relationships between particular events in the schedule and the day as a whole.

The School Day Post a schedule for each school day. Identify important events (start of school, math, music, recess, reading, lunch) using pictures or symbols and times. Include both analog (clock face) and digital (10:15) representations. Discuss the daily schedule each day with students using words such as *before* math, *after* recess, *during* the morning, *at the end of* the school day. Later in the school year you can begin to identify the times that events occur as a way of bridging the general idea of sequential events and the actual time of day.

The Weekend Day Students can create a daily schedule, similar to the class schedule, for their weekend days. Initially they might make a "timeline" of their day, putting events in sequential order. Later in the year they might make another schedule where they indicate the approximate time of day that events occur.

Weather

Keeping track of the weather engages young students in a real-life data collection experience in which the data they collect changes over time. By displaying this ongoing collection of data in one growing representation, students can compare changes in weather across days, weeks, and months, and observe trends in weather patterns, many of which correspond to the seasons of the year.

Monthly Weather Data With the students, choose a number of weather categories (which will depend on your climate); they might include sunny, cloudy, partly cloudy, rainy, windy, and snowy.

If you vary the type of representation you use to collect monthly data, students get a chance to see how similar information can be communicated in different ways. On the following page you'll see some ways of representing data that first grade teachers have used.

At the end of each month (and periodically throughout the month), ask questions to help students analyze the data they are collecting.

About Classroom Routines

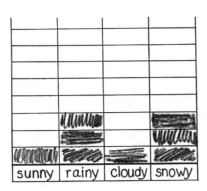

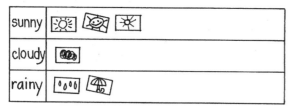

Weather data can be collected on displays like these. In the second example, a student draws each day's weather on an index card to add to the graph. The third example uses stick-on dots.

What is this graph about?

What does this graph tell us about the weather this month (so far)?

What type of weather did we have for the most days? What type of weather did we hardly ever have?

How is the weather this month different from the weather last month? What are you looking at on the graph to help you figure that out?

How do you think the weather graph for next month will look?

Yearly Weather Data If you collect and analyze weather data for some period of time, consider extending this over the entire school year. Save your monthly weather graphs, and periodically look back to see and discuss the changes over longer periods of time.

Another approach over the entire year is to prepare 10-by-10 grids from 1-inch graph paper, making one grid for each weather category your class has chosen. Post the grids, labeled with the identifying weather word. Each day, a student records the weather by marking off one square on one or more grids; that is, on a sunny day, the student marks a square on the "sunny" grid, and if it's also windy, he or she marks the "windy" grid, too.

From time to time, students can calculate the total number of days in a certain category by counting the squares. Because these are arranged in a 10-by-10 grid, some students may use the rows of 10 to help them calculate the total number of days. ("That's 10, and another 10 is 20, and 21, 22, 23.")

Making Weather Representations After students have had some experience collecting and recording data in the grade 1 curriculum (especially in *Survey Questions and Secret Rules*), they can make their own representation of the weather data. For one month, record the weather data on a piece of chart paper (or directly on your monthly calendar), without organizing it by category. At the end of the month, ask students to total the number of sunny days, rainy days, and so forth, and post this information (perhaps as a tally). Students then make their own representation of the data, using pictures, numbers, words, or a combination of these. Encourage them to use clear categories and show the number of days in each.

VOCABULARY SUPPORT FOR SECOND-LANGUAGE LEARNERS

The following activities will help ensure that this unit is comprehensible to students who are acquiring English as a second language. The suggested approach is based on *The Natural Approach: Language Acquisition in the Classroom* by Stephen D. Krashen and Tracy D. Terrell (Alemany Press, 1983). The intent is for second-language learners to acquire new vocabulary in an active, meaningful context.

Note that *acquiring* a word is different from *learning* a word. Depending on their level of proficiency, students may be able to comprehend a word upon hearing it during an investigation, without being able to say it. Other students may be able to use the word orally, but not read or write it. The goal is to help students naturally acquire targeted vocabulary at their present level of proficiency.

We suggest using these activities just before the related investigations. The activities can also be led by English-proficient students.

Investigation 1

heavy, light, heavier, lighter, same

1. Collect a group of three or four objects of distinctly different weights, and two identical objects. For example, gather books of different weights (an atlas, two of the same reading text, a small paperback), or common classroom objects.

2. Lift one of the heavier objects and with exaggerated gestures indicate that it is *heavy*. Repeat with one that is *light*. Lift the two together and identify which is *heavier* and which is *lighter*.

3. Students take turns hefting the two objects as you ask:

   Which one is heavier?
   Which one is lighter?
   Repeat with other object pairs.

4. Hold the identical objects in both hands and explain that neither is heavier or lighter; they are the *same*. Students take turns testing the two to see if they agree.

5. Divide the students into two teams. Pick an object and challenge the teams to find something in the room that is heavier than your object. Continue for several rounds.

6. Repeat the game, this time with teams looking for objects lighter than the one you identify.

7. As follow-up to the game, ask questions that focus on the vocabulary. For example:

   Was the dictionary *heavier* or *lighter* than the notebook?
   Was the pen *heavier* or *lighter* than the scissors?
   What was *heavier* than the stapler?
   What was *lighter* than the stapler?

Investigation 2

container, empty, fill, holds more, holds less, same amount

1. Collect four containers, including two that are identical, one that is much smaller, and another that is much larger. Identify them as *containers,* and turn each over to demonstrate that they are *empty*.

2. Using interlocking cubes, show students how you can *fill* the containers. Then empty the cubes on the table so you have four piles of cubes, each next to the container they once filled. Point out which container *holds more*, which *holds less*, and which two hold the *same amount*.

3. Take one of these containers and fill it again with cubes. Make available containers of other sizes. Challenge students to choose a container that *holds more* than yours. Encourage them to use cubes and dump out the cubes in the two containers for a visual check.

4. Repeat, asking students to find a container that holds less than yours, and again, a container that holds about the same amount as yours.

size, tall, short, wide, narrow

1. Using the same containers and a few more, explain that you are going to group them by *size*. Gather in one group the taller containers; in another group, the shorter ones. Identify each group, tracing the edges with your finger to identify the dimension (height) being focused on.

2. Present another container and ask students to look at its size and decide which group it belongs in, the *tall* group or the *short* group. Repeat with other containers.

3. Using the same or other containers, call attention to another dimension: width. Trace the outer edge of each with your finger as you identify the container as being either *wide* or *narrow*. Ask students to place the containers into two groups according to this dimension.

4. Randomly select objects from the groups as you ask questions about their dimensions. For example:

   Is this bottle tall or short?
   Is this bottle wide or narrow?

Investigation 3

long, longer, longest, short, shorter, shortest

1. Draw five lines of different lengths on the board, aligning them all at one end. Number them 1–5. For example:

   1 _____
   2 _____
   3 ___
   4 _____
   5 _____

   As you point to each line, describe its length in relation to the other lines. For example:

   Line 4 is longer than line 3.
   Line 1 is shorter than line 4.
   Line 3 is the shortest line.

2. Erase the numbered lines. Ask three students in turn to draw a line on the board, one above the other, aligned at one end. Their lines can be of any length, as long as all three are different lengths. Write each student's name on his or her line.

3. Ask the group to compare these three lines.

Blackline Masters

Family Letter   98

Investigation 1
Student Sheet 1, Which Is Heavier?   99
Student Sheet 2, Something to Weigh   100
Student Sheet 3, Heavier, Lighter, the Same   101
Student Sheet 4, Finding Things That Balance   102

Investigation 2
Student Sheet 5, Which Holds More?   103
Student Sheets 6–9, Block Puzzles A–D   104
Student Sheets 10–11, Block Puzzles E–F   108
Student Sheet 12, Comparing Bottles   110
Student Sheet 13, Two Containers   111
Pattern Block Cutouts   112

Investigation 3
Student Sheet 14, Pencil Comparisons   118
Student Sheet 15, Foot Outlines from Home   119
Student Sheet 16, Measuring with Hands and Feet   120
Student Sheet 17, Shorter Than My Arm   121
Student Sheet 18, Foot Match-Ups   122
Student Sheet 19, How Many Cubes Long?   123
Foot Outlines   124
100 Chart   126

General Resources for the Unit
Practice Pages   127

_____, 19____

Dear Family,

For the next few weeks in math our class will be working on a unit called *Bigger, Taller, Heavier, Smaller*. Your child will be doing different kinds of measuring.

We begin by weighing and balancing familiar objects to help develop a sense of what's *heavier* and *lighter*. The children learn to use a balance to compare two or more objects.

Next, we will explore the amounts that different containers can hold. The children fill containers of all shapes and sizes with materials such as water, sand, and cubes to find which containers hold more.

Finally, we'll investigate the *length* of things in our classroom. The children will use their hands and feet, and also cubes, to measure and compare the lengths of various objects.

Measuring experiences at home will support your child's work at school. Here are some ways you can help:

- ■ To talk about weight, collect a small group of objects. Work together to figure out which one is heaviest and which is lightest by holding them in your hands. Since we can't "see" weight, it's important for young children to have lots of experiences *feeling* the weights of things. Ask your child to help you weigh fruit at the grocery store, or figure out which package is the heaviest and will need the most postage.

- ■ As you are cooking and baking, ask your child to help with filling, measuring, and leveling off measuring cups and spoons. You can also do experiments at home, comparing the capacities of different containers. Will the glass hold more water or will the mug? How can you prove which holds more?

- ■ Ask your child to help you estimate length in practical terms. For example, how many chairs can fit along one side of a table? How many steps does it take to walk from the kitchen to the front door?

Although children at this age are not yet learning to use standard units (such as centimeters or ounces), they are very curious about this aspect of measuring. As you use measuring tools (scales, measuring cups, or rulers) around the house, talk to them about how the resulting numbers help you answer your questions. This kind of experience will improve their sense of what measuring is all about.

Sincerely,

Name _____ Date _____

Student Sheet 1

Which Is Heavier?

_____ or _____

_____ or _____

_____ or _____

_____ or _____

_____ or _____

Name _____ Date _____

Student Sheet 2

Something to Weigh

Please bring from home an object to weigh.

It should weigh less than about half a pound.

It should not be taller or wider than about 6 inches.

What did you decide to bring?

_____

_____

Name _____   Date _____

Student Sheet 3

Heavier, Lighter, the Same

Choose one object to put in the balance every time.

What is it? _____

1. Find one thing heavier.

   What is it? _____

2. Find one thing lighter.

   What is it? _____

3. Find one thing that weighs the same.

   What is it? _____

Draw a picture to show how the balance looked for one of these three cases.

Name _____ Date _____

Student Sheet 4

Finding Things That Balance

Get two bags (paper or plastic). Find some things in your kitchen to put in the bags. Make the two bags weigh about the same.

Hold the bags, one in each hand, the way we did in class. Do they seem to balance?

Draw or write what you put in the two bags.

Here's what I put in the first bag:

Here's what I put in the second bag:

Name _____ Date _____

Student Sheet 5

Which Holds More?

We used container _____ and container _____ .

Which container holds more? _____

Draw a picture of the container that holds more.

How did you figure it out? _____
_____

We used container _____ and container _____ .

Which container holds more? _____

Draw a picture of the container that holds more.

How did you figure it out? _____
_____

Name _____ Date _____

Student Sheet 6

Block Puzzle A

Which set of shapes fills the puzzle exactly? Try them both.

Which set of shapes exactly fills Puzzle A? _____

Set 1

Shape	⬡	⏢	▱	◻	▱	△
How many?	0	0	0	9	0	0

Set 2

Shape	⬡	⏢	▱	◻	▱	△
How many?	0	0	3	3	6	2

Optional Show how the blocks fit. Glue on paper shapes or color them in.

Investigation 2 • Sessions 2–4
Bigger, Taller, Heavier, Smaller

Name _____ Date _____

Student Sheet 7

Block Puzzle B

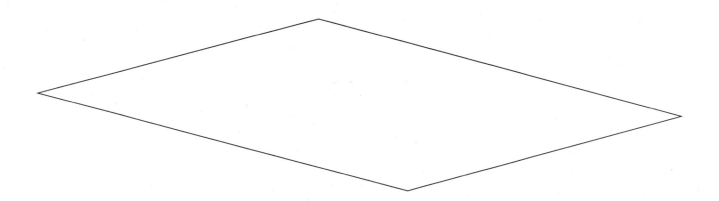

Which set of shapes fills the puzzle exactly? Try them both.

Which set of shapes exactly fills Puzzle B? _____

Set 1

Shape	⬡	⏢	▱	☐	╱╱	△
How many?	0	0	4	0	4	1

Set 2

Shape	⬡	⏢	▱	☐	╱╱	△
How many?	0	0	0	0	12	0

Optional Show how the blocks fit. Glue on paper shapes or color them in.

105

Investigation 2 • Sessions 2–4
Bigger, Taller, Heavier, Smaller

Name _____ Date _____

Student Sheet 8

Block Puzzle C

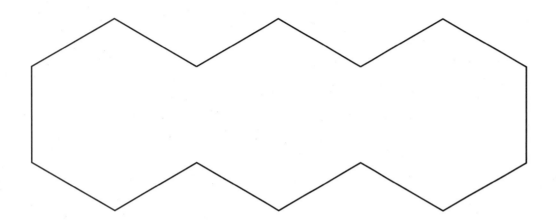

Which set of shapes fills the puzzle exactly? Try them both.

Which set of shapes exactly fills Puzzle C? _____

Set 1

Shape	⬡	⏢	▱	☐	╱╱	△
How many?	0	2	3	0	0	3

Set 2

Shape	⬡	⏢	▱	☐	╱╱	△
How many?	0	2	3	0	0	6

Optional Show how the blocks fit. Glue on paper shapes or color them in.

Name _____ Date _____

Student Sheet 9

Block Puzzle D

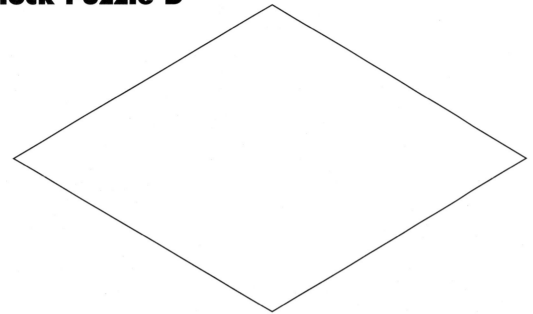

Which set of shapes fills the puzzle exactly? Try them both.

Which set of shapes exactly fills Puzzle D? _____

Set 1

Shape	⬡	⏢	▱	□	╱╱	△
How many?	2	2	6	0	0	2

Set 2

Shape	⬡	⏢	▱	□	╱╱	△
How many?	0	4	10	0	0	2

Optional Show how the blocks fit. Glue on paper shapes or color them in.

Name _____ Date _____

Student Sheet 10

Block Puzzle E

Make your own puzzle.
Fill the outline.

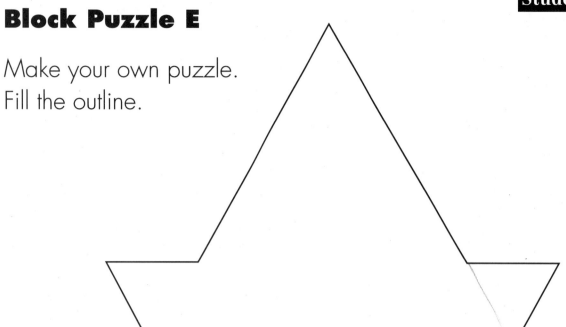

Write down how many blocks of each kind you used. Fill in the other chart with a set of blocks that does **not** solve the puzzle.

Set 1

Shape	⬡	⬠	◇	□	▱	△
How many?						

Set 2

Shape	⬡	⬠	◇	□	▱	△
How many?						

Investigation 2 • Sessions 2–4
Bigger, Taller, Heavier, Smaller

Name _____ Date _____

Student Sheet 11

Block Puzzle F

Make your own puzzle.
Fill the outline.

Write down how many blocks of each kind you used. Fill in the other chart with a set of blocks that does **not** solve the puzzle.

Set 1

Shape	⬡	⬟	◇	□	╱╱	△
How many?						

Set 2

Shape	⬡	⬟	◇	□	╱╱	△
How many?						

Investigation 2 • Sessions 2–4
Bigger, Taller, Heavier, Smaller

Name _____ Date _____

Student Sheet 12

Comparing Bottles

Use this sheet when you compare 5 bottles to see how much water they hold.

Keeping track:

Which two bottles hold the same amount of water?

bottle _____ and bottle _____

Draw these two bottles here.

Investigation 2 • Sessions 5–7
Bigger, Taller, Heavier, Smaller

Name _____ Date _____

Student Sheet 13

Two Containers

Choose two empty containers. Find out which holds more water.

You can use something to help you fill them, such as a measuring cup, a spoon, or a small paper cup.

Draw a picture of the two containers you used:

> **Note to Families**
> Instead of water, you can provide something dry to measure with, such as sand, rice, cat litter, dried beans. Find a place for your child to work, where spills don't matter.

Write what you did and what you found out.

_____

_____

_____

Investigation 2 • Sessions 5–7
Bigger, Taller, Heavier, Smaller

PATTERN BLOCK CUTOUTS (page 1 of 6)

Duplicate these hexagons on yellow paper and cut apart.

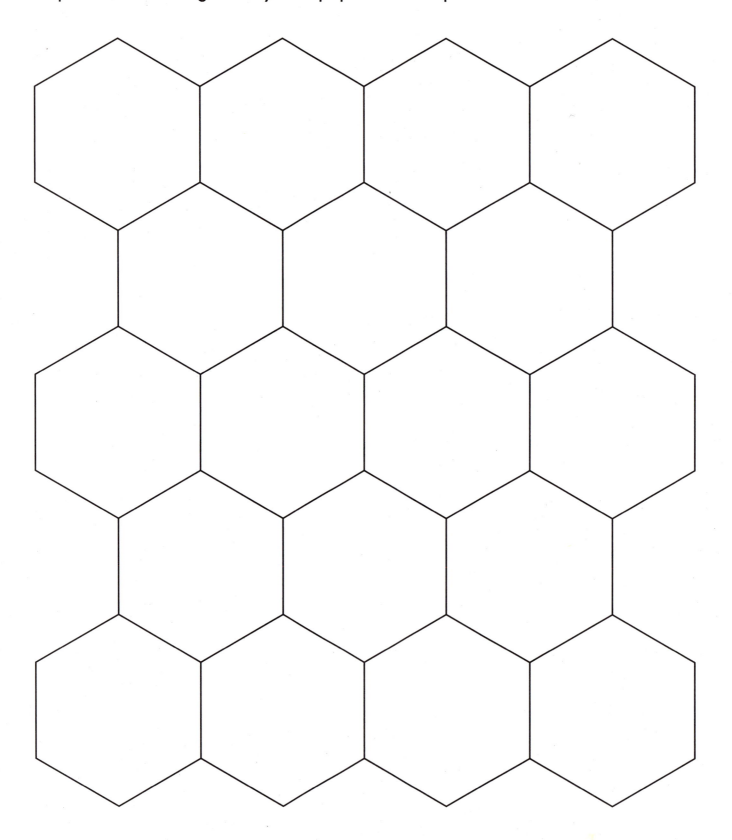

PATTERN BLOCK CUTOUTS (page 2 of 6)

Duplicate these trapezoids on red paper and cut apart.

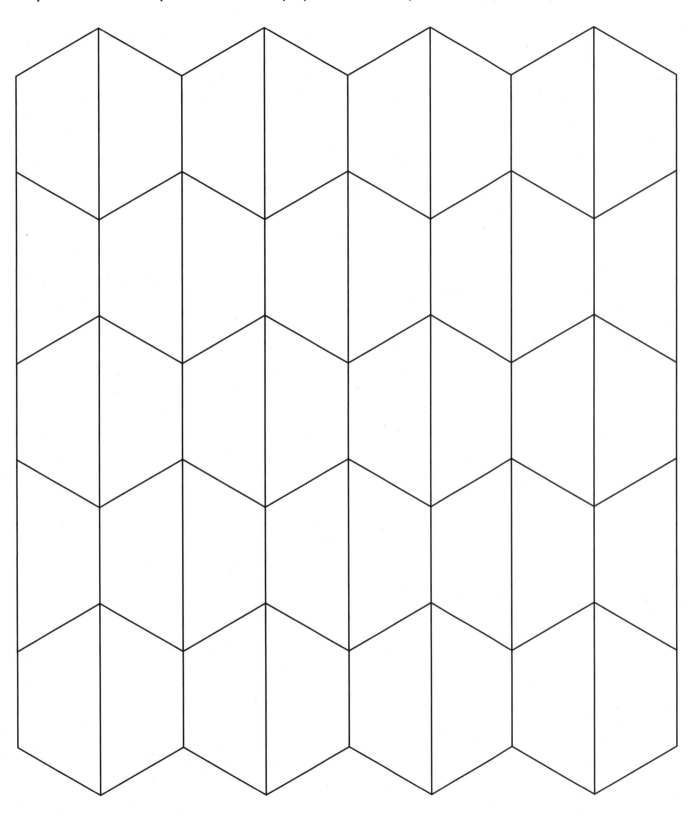

Investigation 2 Resource
Bigger, Taller, Heavier, Smaller

PATTERN BLOCK CUTOUTS (page 3 of 6)

Duplicate these triangles on green paper and cut apart.

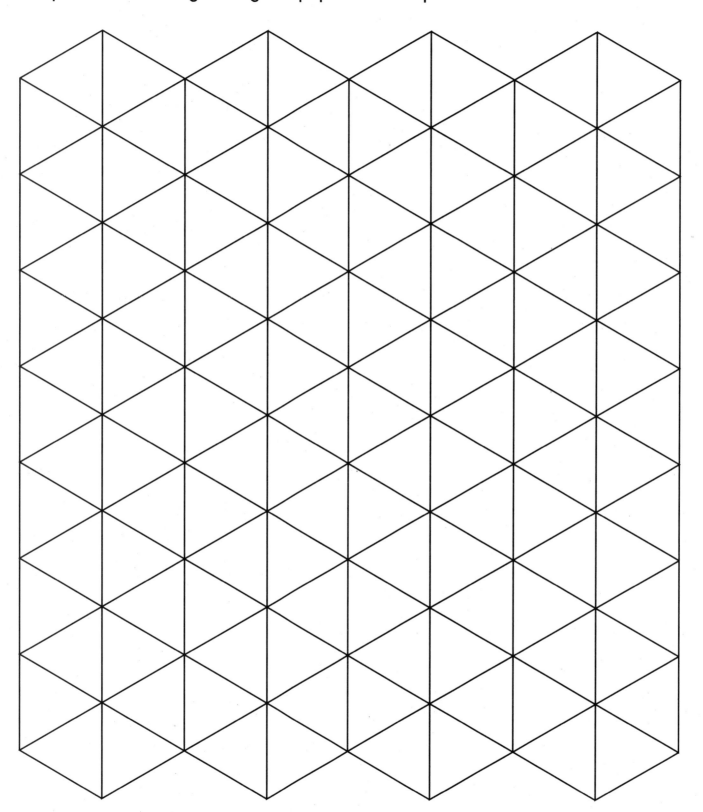

PATTERN BLOCK CUTOUTS (page 4 of 6)
Duplicate these squares on orange paper and cut apart.

Investigation 2 Resource
Bigger, Taller, Heavier, Smaller

PATTERN BLOCK CUTOUTS (page 5 of 6)

Duplicate these rhombuses on blue paper and cut apart.

PATTERN BLOCK CUTOUTS (page 6 of 6)
Duplicate these rhombuses on tan paper and cut apart.

Name _____ Date _____

Student Sheet 14

Pencil Comparisons

In each box, write or draw something to compare with your pencil. Which one is longer? Circle it. If they are the same size, circle both.

my pencil	my pencil
my pencil	my pencil
my pencil	my pencil
my pencil	my pencil

© Dale Seymour Publications®

Investigation 3 • Session 1
Bigger, Taller, Heavier, Smaller

Name
Date

Student Sheet 15

Foot Outlines from Home

As we study length, we are going to compare the lengths of different feet.

You need to trace outlines of at least four different feet. Bring your foot outlines back to school.

- Use a new sheet of paper for each foot. (Use any paper, even newspaper.)

- Use a pencil, crayon, or marker.

- Trace each foot with socks on, no shoes.

- Trace only one foot for each person. It doesn't matter which foot.

- Label each foot with the person's name.

- Cut out each outline.

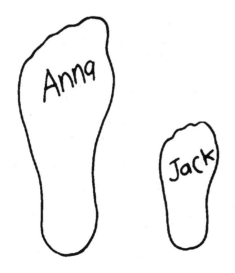

Name _____   Date _____

Student Sheet 16

Measuring with Hands and Feet

Find things to measure. What did you find?
Draw each object or write its name. How long is it?
Write its length. Circle the word (**hands** or **feet**)
that tells what you used for measuring.

Thing measured	Length	Measured with
_____	_____	hands feet
_____	_____	hands feet
_____	_____	hands feet
_____	_____	hands feet
_____	_____	hands feet
_____	_____	hands feet
_____	_____	hands feet
_____	_____	hands feet

© Dale Seymour Publications®

Investigation 3 • Sessions 2–3
Bigger, Taller, Heavier, Smaller

Name _____ Date _____

Student Sheet 17

Shorter Than My Arm

Bring in two things from home to measure. They should be nonbreakable. Each thing should be shorter than your arm.

We will measure these things in math class.

What did you decide to bring?

_____

_____

Investigation 3 • Session 2
Bigger, Taller, Heavier, Smaller

Name _____ Date _____

Student Sheet 18

Foot Match-Ups

Foot length   Whose foot is a match?

about 6 cubes long   _____

about 9 cubes long   _____

about 12 cubes long   _____

about 14 cubes long   _____

less than 6 cubes long   _____

more than 14 cubes long   _____

Investigation 3 • Session 3
Bigger, Taller, Heavier, Smaller

Name _____ Date _____

Student Sheet 19

How Many Cubes Long?

Name of object   Picture   How long

_____   _____   _____ cubes

_____   _____   _____ cubes

_____   _____   _____ cubes

_____   _____   _____ cubes

_____   _____   _____ cubes

Which is your longest object? _____

Which is your shortest object? _____

© Dale Seymour Publications®

Investigation 3 • Sessions 4–5
Bigger, Taller, Heavier, Smaller

FOOT OUTLINES (page 1 of 2)

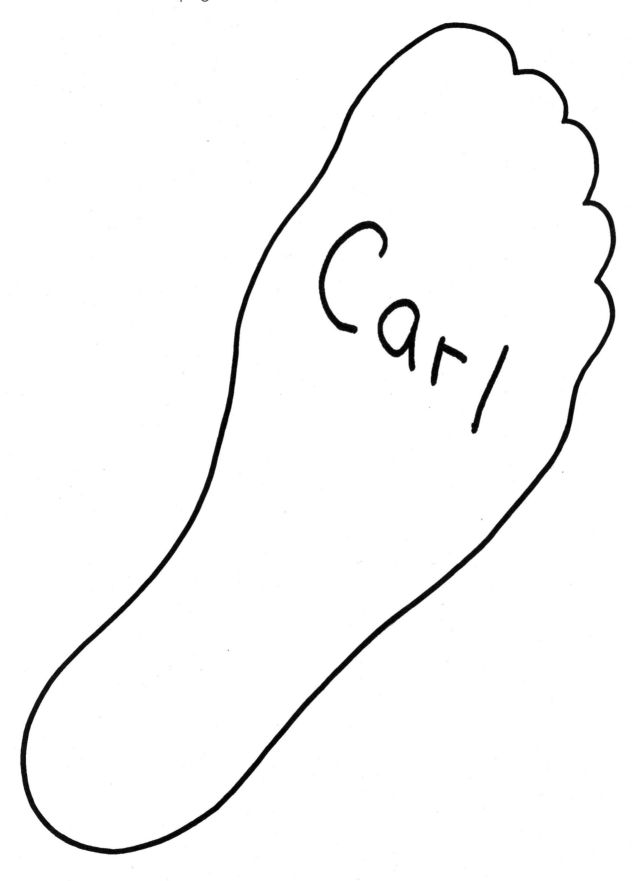

FOOT OUTLINES (page 2 of 2)

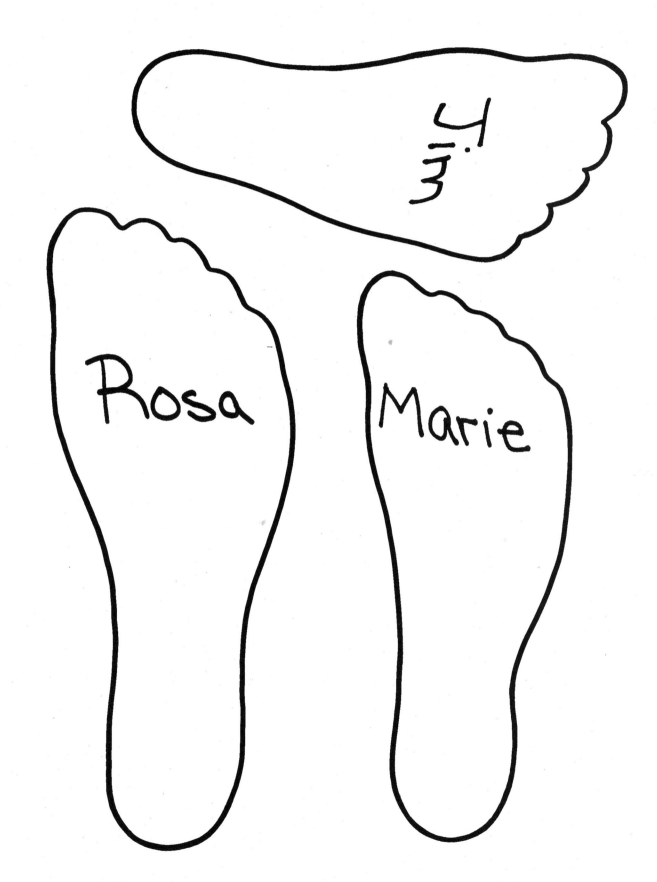

100 CHART

1	2	3	4	5	6	7	8	9	10
11	12	13	14	15	16	17	18	19	20
21	22	23	24	25	26	27	28	29	30
31	32	33	34	35	36	37	38	39	40
41	42	43	44	45	46	47	48	49	50
51	52	53	54	55	56	57	58	59	60
61	62	63	64	65	66	67	68	69	70
71	72	73	74	75	76	77	78	79	80
81	82	83	84	85	86	87	88	89	90
91	92	93	94	95	96	97	98	99	100

Practice Pages

This section provides optional homework for teachers who want or need to give more homework than is suggested to accompany the activities in this unit. With the games and problems included here, students get additional practice in learning about number relationships and solving number problems. Whether or not the *Investigations* unit you are presenting in class focuses on number skills, continued work at home on developing number sense will benefit students. In this unit, optional practice pages include the following:

Ten Turns This game is introduced in the grade 1 unit *Building Number Sense*. If your students are familiar with the game, simply send home the directions and the Ten Turns Game Sheet. If your students have not played this game before, you will probably want to introduce it in class and help students play once or twice before sending it home. If they are very familiar with the game, you might suggest one of the listed variations.

Number of the Day For this activity, choose as the "number of the day" a number you want students to practice with, such as 10 or 12, or perhaps today's date. Students write expressions that are equal to that number. For example, if the number of the day is 12, possible expressions include these:

$$10 + 2 \qquad 3 + 7 + 2 \qquad 4 + 4 + 4 \qquad 13 - 1$$

Introduce the activity in class to be sure that students understand the range of possible responses. Practice Page A can be used more than once; fill in a number before copying it each time you want students to work on today's number at home.

Story Problems Students work with story problems in the grade 1 units *Building Number Sense* and *Number Games and Story Problems*. The problems included here for practice specifically continue the type of work students did in those units, solving problems and recording their solutions using pictures, numbers, and words. If you haven't done either of these units in class, it would be better to assign other practice pages for homework. For additional practice, you can make up other problems in this format, using numbers and contexts that are appropriate for your students.

Name _____ Date _____

Ten Turns

Materials: One number cube
Counters (50–60)
Ten Turns Game Sheet

Players: 2

Object: With a partner, collect as many counters as you can.

How to Play

1. Roll the number cube. What number did you roll? Take that many counters to start your collection. Write the number you rolled and the total number you have. (For the first turn, these numbers are the same.)

2. On each turn, roll the number cube and take that many counters. Find the total number of counters you and your partner have together.

3. After each turn, write the number you rolled and the new total.

4. Play for 10 turns.

Variations

a. Play for fewer turns or more turns.

b. Roll two number cubes on each turn.

c. Instead of a number cube, use the Number Cards for 1 to 6. Mix them and turn up one at a time.

Note to Families

For counters, use buttons, pennies, paper clips, beans, or toothpicks. If you don't have a number cube or number cards, use slips of paper numbered 1–6. If you don't have the Ten Turns Game Sheet, keep track of the numbers for each turn and the new total on a blank sheet of paper.

Practice Page

Bigger, Taller, Heavier, Smaller

Name _____   Date _____

Ten Turns Game Sheet

Turn 1.   I rolled _____.   Now we have _____.

Turn 2.   I rolled _____.   Now we have _____.

Turn 3.   I rolled _____.   Now we have _____.

Turn 4.   I rolled _____.   Now we have _____.

Turn 5.   I rolled _____.   Now we have _____.

Turn 6.   I rolled _____.   Now we have _____.

Turn 7.   I rolled _____.   Now we have _____.

Turn 8.   I rolled _____.   Now we have _____.

Turn 9.   I rolled _____.   Now we have _____.

Turn 10. I rolled _____.   Now we have _____.

Practice Page
Bigger, Taller, Heavier, Smaller

Name _____   Date _____

Practice Page A

The number of the day is _____

Write equations that show ways to make the number of the day.

Name _____ Date _____

Practice Page B

Nina has 17 yellow marbles.
She also has 5 green marbles.
How many marbles does she have?

Show how you solved this problem.
Use pictures, numbers, or words.

Name _____   Date _____

Practice Page C

Rico had 21 crackers. He ate 8 of them. How many crackers did he have left?

Show how you solved this problem. Use pictures, numbers, or words.

Practice Page D

Eric picked 19 flowers. He gave
12 flowers to his mother.
How many flowers did he have left?

Show how you solved this problem.
Use pictures, numbers, or words.

Name _____   Date _____

Practice Page E

Lexi collects stickers. She has 9 stars, 10 dinosaurs, and 6 hearts. How many stickers does she have?

Show how you solved this problem. Use pictures, numbers, or words.